Olivier C. Rieppel

Fundamentals of Comparative Biology

1988
Birkhäuser Verlag
Basel · Boston · Berlin

Author's address:
Dr. Olivier C. Rieppel
Paläontologisches Institut
und Museum der Universität
Künstlergasse 16
CH-8006 Zürich
Switzerland

Library of Congress Cataloging-in-Publication Data

Rieppel, Olivier
   Fundamentals of comparative biology / Olivier C. Rieppel.
      p. cm.
   Includes index.
   ISBN 08176-1956-9 (U.S.)
   1. Evolution. 2. Biology. I. Title
QH368.2.R54 1988   88-22253
574--dc 19

CIP-Titelaufnahme der Deutschen Bibliothek

Rieppel, Olivier:
Fundamentals of comparative biology / Olivier C. Rieppel. -
Basel ; Boston ; Berlin : Birkhäuser, 1988
   ISBN 3-7643-1956-9 (Basel ...) brosch.
   ISBN 0-8176-1956-9 (Boston) brosch.

© 1988 Birkhäuser Verlag
   P.O. Box 133
   CH-4010 Basel / Switzerland

Printed in Germany
ISBN 3-7643-1956-9
ISBN 0-8176-1956-9

CONTENTS

CONTENTS

INTRODUCTION

Comparative biology has experienced a revival of interest during the last few years, a movement which is not only due to the availability of new and sophisticated experimental techniques in functional morphology, but also and even more so to a renaissance in systematics and evolutionary biology. The debate on tempo and mode of evolution has highlighted a number of problem areas which fall into the domain of classical comparative morphology. Modern appreciation of long-standing issues has focused on a series of controversial concepts which may be presented as three basic antitheses of comparative biology.

HOLISM	——	**ATOMISM**
FORM	——	**FUNCTION**
HIERARCHY	——	**CONTINUITY**

It is my purpose, in this book, to evaluate these three antitheses of comparative biology in philosophical, historical and methodological focus. Firstly I shall discuss some philosophical premises which underlie the search for order in nature and its causal explanation by laws of ontogeny and phylogeny. Secondly I will outline the rise of different 'ways of seeing' in the course of the history of comparative biology, and the persistence of some basic issues into modern debates. I will reach the conclusion that different philosophical premises and different methodological procedures result in alternative conceptualizations of ontogeny and phylogeny which are complementary to each other. The whole argument therefore is one for pluralism in modern biology.

Holism versus Atomism

Holism views the organism as a developmentally and functionally integrated whole; it emphasizes the correlation of structures and the functional interdependence of organ systems. It was taken to its logical extreme in the preformationism of the 18th century naturalist and philospher Charles Bonnet: the "Tout Organique" develops from an organized fluid which must contain all its preformed parts - if not in a visible, then certainly in a functional condition. According to Bonnet, this is rendered necessary by the developmental and functional correlation of parts, since after all: development presupposes nutrition, which in turn requires a circulatory system, which depends on a pumping heart, presupposing the brain for innervation, etc. Holism is essentially a-historical: the stability of structure is emphasized, and if change is to occur, it must somehow affect the whole organization in a harmonious manner.

Atomism in contrast is related to a reductionist perspective: the organism is reduced to or decomposed into its constituent elements, the structure and function of which is analyzed in isolation. Change of organization is viewed as the result of a changing structure or function of fundamentally variable and interchangeable parts. Atomism is essentially

Darwinian, or, to put it into the correct historical perspective: Darwinism is based on an atomistic background. The organism is viewed as being composed of an aggregation of fundamentally variable constitutional elements.

Form versus Function

Structuralism (Hughes and Lambert, 1984) presupposes the primacy of form over function. It is form which determines function. A causal explanation of the regularity and invariance of form must therefore relate to lawful generative mechanisms of structure, i.e. to the integration of developmental processes and constraints. As Shubin and Alberch (1986: 377) have put it: "...The quest for a general set of principles of form is legitimate if we were to exchange the metaphysical concept of the *Bauplan* for a mechanistic one based on principles of morphogenesis and internal integration".

Functionalism (Lambert and Hughes, 1984), on the other hand, emphasizes the primacy of function over form. It is function which determines form, and the causal explanation for the change of form is provided by the theory of natural selection. "Nature may be compared to a surface covered with ten-thousand sharp wedges...", as Charles Darwin put it, and it is the action of these wedges which molds the form so as to assure the adaptation of organisms.

Form has been analyzed in a theoretical morphospace by Martin J.S. Rudwick, who found empty spaces but also lack of optimization of form in relation to function. These realms of "forbidden morphologies" as well as compromise organization have called into question the sufficiency of the functionalist approach as an explanative paradigm by demonstrating that causal factors must be allowed for, acting from "within" rather than from "without". Form is not only molded by natural selection, but its creation during ontogeny is subject to developmental constraints. In addition to historical and genetic constraints, A. Seilacher has identified "material constraints" (*bautechnische Zwänge*) which are the subject of biophysics.

Hierarchy versus Continuity

Holism, or the rejection of the reductionist way of decomposing organic structure into its constituent elements to arrive at a Cartesian machine, has been likened by Lauder (1982: xvi) to a hierarchical view of organization.

The hierarchical view of organization emphasizes the distinctiveness of subordinated levels of complexity of structure and function, each with emergent and irreducible properties of its own, but also connected to the levels above and below by upward and downward causation. Any general theory of comparative biology must integrate this subordinated hierarchy of nature into a synthetic perspective without distortion which would result from a reduction of explanatory theories to inappropriate levels of complexity. The holistic or hierarchical perspective therefore emphasizes the complexity and individuality of subordinated levels of organization and thence discontinuity.

In constrast, the reductionist approach explains the properties of higher levels of complexity by reduction to lower and less complicated levels of the hierarchy. As emergent properties drop out of focus, the hierarchically organized levels of complexity lose their distinctiveness. Reduction to a common denominator permits ready comparison and therewith the construction of continuous transformational series which are explained by an unbroken nexus of upward causation leading from "molecules to men".

4

The hierarchical approach is static: it emphasizes the stability of structure and the difficulty of change (S.J. Gould). The atomistic perspective on the other hand is dynamic, permitting continuous change from the elementary to the complex by the reduction of the complex to a composition of elementary parts.

Antithesis or Synthesis?

Three basic antitheses of comparative biology have been identified: Holism, structuralism and the hierarchical view of nature contrasts with reductionism, functionalism and a Darwinian theory of change. These three antitheses are all related to one of the most fundamental principles of natural science, the Leibnizian principle of continuity. In a somewhat simplistic manner it could be argued that the contrasting views result from the acceptance or rejection of this principle as a cornerstone of comparative biology. Continuity of pattern and process was a metaphysical conception of the deductivism of Leibniz; its acceptance or rejection in contemporary biology results in two fundamentally different "ways of seeing" (Hughes and Lambert, 1984; Rieppel, 1987a).

It is the aim of the present book to argue for the complementarity of these alternative "ways of seeing". Neither perspective is in itself sufficient to produce a complete explanation of natural phenomena, nor is it possible to reduce one perspective to the other. Observation and explanation may proceed from either point of view, resulting in different appearances subject to alternative explanatory theories. There result alternative and complementary views of a whole which as such remains incomprehensible.

It is appropriate at this point to express my gratitude to colleagues, friends and students who, by discussion as much as by their questions or criticisms, have helped me to develop my ideas. In particular, I should like to thank Dr. C. Patterson, London, and Dr. A.L. Panchen, Newcastle-upon-Tyne, as well as two anonymous reviewers who all read the manuscript and offered many invaluable comments and suggestions improving the text. Thanks are also due to Dr. A. Bally, Birkhäuser Verlag, who provided technical assistance as well as encouragement and moral support throughout this project.

CHAPTER 1

A PHILOSOPHER's VOCABULARY

The foundation of comparative biology goes back to the work of Aristotle; ever since, this science has preserved close links to philosophy. In 1824, Etienne Geoffroy Saint-Hilaire used the term "*philosophication*" to characterize his search for homologies, while his colleague Etienne Serres published a series of monographs dealing with "transcendental anatomy" during the first half of the nineteenth century. His views came close to the ideas of the proponents of *Naturphilosophie*, popular in Germany and in pre-Darwinian England. Some modern approaches to comparative biology such as "transformed" or "pattern" cladism and the structuralist "way of seeing" to be discussed in the course of this book have been accused of falling back on "idealistic morphology". This makes it appear useful, if not necessary, to introduce the philosophical vocabulary used in comparative biology to assure proper understanding of these theory-laden notions.

Plato: "Being" and "Becoming"

Timaios was, at the dawn of the Middle Ages, the only work known of Plato, and it greatly influenced the development of Western thought, particularly in the context of developing Christian tradition. In this treatise, Plato discussed the Creation of the world, distinguishing between the categories of 'being' and 'becoming', categories which preserved an eminent significance in the Neoplatonic philosophy dominating the views of many Church Fathers (Gilson, 1955).

The use of the Platonic genera of 'being' and 'becoming' in biology was related to Natural Theology. The question, derived from Paulinian theology and already raised by the Church Father Aurelius Augustinus, was how an eternal and therefore immutable being like God could suddenly decide to create a world in time and space without being subject to change himself. Change, if only a change of motivation, would throw the Divine being into a time-axis. The answer to this question was based on the Platonic distinction of the genera of 'being' and 'becoming'. The divine ideas, the Plan of Creation according to Natural Theology, is a category of 'being', an ideal category of thought lying beyond the constraints of time, space and therewith of causality; it is eternal and consequently immutable. Both for Plato, and for proponents of Natural Theology, this was the realm of ultimate reality, inaccessible to human cognition but effective as a pattern upon which the world was created. Creation therefore corresponds to the actualization or materialization of the divine ideas, 'being' is turned into 'becoming' under the constraints of time, space, and therewith causality. This is how God created the world through time, effecting change which did not affect him: his eternal ideas became actualized, i.e. materialized.

The Platonic distinction of the categories of 'being' and 'becoming' aimed at the distinction of logical necessity from historical contingency. The ever-changing appearances of the material world were not considered to be subject to the reign of universal laws in the same sense as are logical relations. The relations of numbers are universally true, and hence were considered godlike by the Pythagoreans. Platonic idealism is related to a Pythagorean background, geometry was chosen as paradigm for the universal truth of lawful relations: the sum of the angles in a triangle is always 180° in Euklidean space, at any time, anywhere in the universe, and for anybody who cares to examine the truth of this statement. The same, it was felt, would not hold for historical contingencies.

6

The concept of universal laws was related to essentialism. Scholastic philosophy viewed the essence of things as a transcendental property, imparting objective identity on ever-changing material appearances. As an element of 'being' and hence immutable, the essence determined the real nature of things. What from the modern perspective appears to be an idealistic concept constituted the realism of scholastics: only by virtue of its essence could a material appearance become the object of a universal law and therewith attain the status of reality, created according to the pattern of a divine idea. Lawful relations, however, could only be universally true if the essence remained immutable and thus transcended the properties of ever-changing matter.

Materialism and reductionism have banned Platonism and therewith transcendental properties from natural sciences, but this does not necessarily result in an elimination of essentialism. It is true that the *"Typus"* or *"Bauplan"* of idealistic philosophy constituted an ideal, abstract and therefore immutable essence, reducing the ever-changing multiplicity of organic appearances to the static and immutable unity of type, but this notion was not always linked to Creationism. It could also be attributed to "nature" viewed in a vitalistic perspective. Aristotle, for instance, attributed the essence of material appearances not to the realm of divine thought, but to matter itself in terms of a "soul" which, according to him, represented the "principle of knowledge" of form: the form of an organism is 'becoming', being actualized during ontogeny under the direction of the immortal soul, i.e. according to the everlasting "principle of knowledge". The organism is always changing, being born, growing up, reaching adulthood and decaying after death. However, by virtue of its being a member of a species it partakes in eternity. The organism itself is nothing but an ephemeral appearance; the species-specific form on the other hand belongs to the category of 'being', ever remaining constant and immutable, as it is actualized under the guidance of the immortal soul.

Even beyond vitalism it is possible to defend essentialism in a modern sense, understanding the essence as a contingent property of matter (Rosenberg, 1985), imparting identity to an appearance which itself is subject to continuous change in all its other attributes. Modern structuralism, for instance, views the *"Bauplan"* as an empirical concept, the constancy of structural patterns resulting from constrained mechanisms of ontogenesis (Shubin and Alberch, 1986). These mechanisms of ontogenesis have been designated as the essence of a taxon by Webster (1984): it represents an element of 'being' in a continuously 'becoming' organism.

William Ockham and the Hypothesis of the "Genius Malignus"

Natural science, if aiming at the discovery of lawful relations permitting the prediction of future events, depends on the uniformity and reliability of natural laws, conceived of as 'secondary causes' in Natural Theology. The method of natural science is thrown into question by the conception of God as almighty, free to intervene in Creation, superseding natural laws and causing miracles as well as catastrophies. Direct interventions of God in the natural order of things are the 'primary causes' of Natural Theology.

Even worse, God might be hypothesized as a *genius malignus*, creating illusions in man who depends on objective perception and cognition in order to deduce natural laws from observation! This hypothesis was the starting point of William Ockham's (*ca.* 1280-1349) argument, distinguishing the act of perception from the judgement as to whether the object perceived does indeed have material reality. If God is able to create in man the illusion of perceptions, the material world is no longer necessarily and immediately accessible to human cognition. Although it may reasonably be assumed that in the natural order of things a material and hence "real" object is the efficient cause of perception, this needs not necessarily be so, and all knowledge must therefore remain hypothetical. Science is reduced to a system of notions designating material objects the "real" existence of which must be hypothesized (Imbach, 1981).

As the material world eludes immediate perception, all knowledge must remain hypothetical: the hope for objectivity and truth must be abandoned. However, any observation can be explained on more than one theory, of which any or none may be correct, i.e. "true". The scientific method must therefore include a principle permitting the evaluation of competing theories. Ockham introduced the *principle of parsimony* to this end. Of several possible solutions, the most economical one is to be accepted. In Ockham's terms this meant that science should avoid notions which cannot be based on empirical evidence, which cannot be deduced from self-evident principles, or which are not guaranteed by Revelation (Copleston, 1976). For the modern reader, K.R. Popper (1976a) put it as follows: an indeterminate number of potential curves can be drawn to link up a number of fixed points scattered in three dimensions. The parsimony principle dictates that the most economical among the many possible curves be chosen, i.e. the shortest.

Ockhams's philosophy also served as starting point of the development of nominalism, a philosophical doctrine opposed to realism (essentialism). It will be recalled that realists ascribed essences to material appearances which determine their reality. Objects sharing the same essence become members of an Aristotelian class which is believed to have a real existence. Aristotle and his medieval disciples believed that the members of a species share in a common essence, i.e. its soul; the species therefore constituted an Aristotelian class.

Under Ockham's premises, the essences dissolve into mental abstractions. The only entities of unquestionable existence are the tangible individuals. The abstraction of universals (shared attributes) from individuals, which permits their classification within a common class, designated by a common notion, corresponds to an act of ordering intelligence. Notions and classes have no material reality - or at least such cannot be ascertained - but are the abstract product of human cognition, ordering appearances according to invariables which are perceived but which have no claim to reality.

Aristotelians define a class by its essence: it is the definition of a reality. Nominalists claim that universals define a name only, i.e. the notion designating a class with no real existence.

Galilei and Newton: Secondary Causes and Natural Laws

In the shadow of Platonism, science strove for the discovery of eternal truth, formulated in terms of natural laws. Plato would accept absolute truth only with respect to his category of 'being' which alone is eternal and hence immutable, i.e. comprehensible in terms of ever-lasting laws. Material phenomena are 'becoming', subject to continuous change and thus eluding the relations of eternal truth.

Galilei (1564-1642) paved the way for a new understanding of natural laws. Planetary motions, it is true, are an attribute of the world of 'becoming', since motion implies time and space. But the astronomer recognized that planetary movement constantly reproduces itself according to a uniform pattern, which can be expressed in mathematical terms. From this observation emerges the possibility to understand the material appearance of ever-changing phenomena as an expression of underlying causes which determine their change, i.e. movement, in a constant and law-like manner (Cassirer, 1969). It is thus the underlying cause of regularity which becomes an element of 'being' in the explanation of natural processes. On the basis of the principle of uniformity, these underlying causes (secondary causes enacted by God) can be formulated as natural laws, expressed in mathematical terms as they determine the dynamic permanence (Regnéll, 1967) of material appearances.

Were these laws "true"; did they relate to "essential" properties of matter? Newton's (1643-1727) conclusion was that these questions might as well be left unanswered. If a constant regularity of planetary motion is observed, if an object falls back on earth each time it is picked up and released, the regularity can be explained on the law of gravity. Whether gravity constitutes an "essential" property of the material world is irrelevant to the question asked and answered. To do science it is enough to search for laws which are based on and predict the regular, i.e. lawful relations of appearances; the aim is to understand the world as it appears to the investigator.

Descartes: The Principle of Uniformity

The significance of the principle of uniformity as prerequisite of the scientific research program is best discussed in the context of Cartesian rationalism, which in turn necessitates the distinction of induction and deduction. Induction basically proceeds from observation to the formulation of explanatory theories, while deduction starts with a hypothesis permitting predictions to be tested by experiment.

The first philosopher who explicitly based the scientific method on the interplay of induction and deduction was Francis Bacon (1561-1626). Science was claimed to proceed by the induction of natural laws (general statements) from the observation of particular processes (particular statements). These laws must be tested by the deduction and subsequent observation of as yet unrecorded processes. The test may result in a revision, falsification or verification of the laws under investigation.

For Bacon, research was not based on a contemplative approach, but on an interactionist approach (Krohn, 1987). Bacon did not concentrate on ontological or epistemological problems, but rather on the design of useful research programs and the necessary experimental methods. His ambition was not the discovery of everlasting truth, but the design of new experiments which have to be judged by their originality and usefulness in the promotion of human welfare. Insofar as the experimental interaction with nature promotes knowledge, this knowledge provides the power for the manipulation of natural processes. The investigator is to ask a well defined question, and to search for its answer by experiment. This is the only way to gain access to knowledge of relevant causations, the future effects of which can be deduced from the laws which have been discovered by induction. Knowledge thus becomes equated with power over natural processes, and science becomes a method for the successful management of the future on the basis of successful predictions of natural processes. It may be stressed that, according to Bacon, the inductive search for lawful regularity starts out from a well-defined question originating from a carefully designed research program. This original hypothesis may be corroborated by induction, or falsified by deduction. In that sense, Bacon may be claimed to represent an early forerunner of falsificationism, which was fully developed in modern times by K.R. Popper (Krohn, 1987:143).

Descartes (1596-1670) on the other hand mistrusted induction as a reliable path to perfect knowledge - and perfect knowledge alone would guarantee the unfailing success of predictions! After all, induction is based on the priority of observation, and everybody knows how easily the perceptive powers of man are led astray by optical illusions. This is why Bacon built his claim for *hypothetical* knowledge on experimentation, i.e. on the interaction with natural phenomena, while Descartes built his claim for *perfect* knowledge on the priority of logical deduction, since logical necessity alone is the hallmark of universal truth.

Descartes' philosophy started out from the self-consciousness of the thinking being: *cogito, ergo sum*. Rationalism is based on deductive reasoning, structured according to the laws of logic and beginning with a number of "immediately" ("intuitively") obvious principles or "innate ideas". The rationalist was building the world out of his mind, as his empiricist critics used to say. Descartes, returning to the hypothesis of the *genius malignus* ("*Dieu trompeur*"), mistrusted induction as a possible path to knowledge. In view of the hypothesis of a possibly cheating God, he had to build a metaphysical bridge (Williams, 1981) to warrant the certainty of knowledge gained by deductive reasoning and its conformity to the reality of the material world: the *veracitas divina*. The benevolent God would not cheat; He would not prevent his Creature, which He endowed with intelligence and the power of logical reasoning as well as with innate ideas, from the cognition of truth. Philosophers have raised the question whether by this profession of faith Descartes got caught in circular reasoning, and whether a fundamental circle can at all be avoided in the quest for absolute truth (Williams, 1981; Röd, 1982), but a discussion of this important point lies beyond the scope of the present topic, even more so since the search for everlasting wisdom cannot be its aim.

Was there any way to check on the validity of the faith in the truthfulness of God? There was, according to Descartes, who emphasized the possibility of a safe and correct prediction of processes as deduced from known natural laws. Prediction, serving the successful management of the future, must presuppose the uniformity of natural laws, and this is given by the fact that in Descartes' view, natural laws were enacted by God, who also guaranteed their continuing validity. God himself is eternal and hence immutable, and the laws enacted by him must correspond to His own nature. God's Almightiness is not restricted by His own submission to His laws, since He enacted them like a monarch, i.e. at free will.

Leibniz: The Principle of Continuity

Like Descartes before him, Leibniz (1646-1716) was a rationalist, basing his theory of cognition on deduction. He, too, had to start out from some basic principles which could themselves no longer be questioned, and one of these principles, in fact the cornerstone of his scientific program, was the principle of continuity. Leibniz stressed this principle in opposition to occasionalism, a philosophical doctrine developed by some of Descartes' successors such as Malebranche.

Above the distinction was made between primary and secondary causes, and it was stressed that the scientific program must depend on the uniform action of secondary causes, while God must be banned from direct intervention in His Creation. This distinction might have proved sufficient for the investigation of physical relations, but not for the investigation of psychological relations, including the problem of the relation of the cognizing mind to the recognized matter.

Descartes, in his *Principia Philosophiae* of 1644, a summary and textbook of his philosphy containing the formulation of his famous laws of motion, had equated matter with extension (*extensio*) as opposed to mind (*cogitatio*) which was without extension. That matter should have extension was prerequisite for Descartes' laws of motion: as two pieces of matter cannot occupy the same space at the same time, they must push and therefore influence each other as they meet while in motion. The question remained, however, how the unextended mind could be affected by matter, for instance in terms of bodily sensations, or conversely, how mind could impose its will on matter, for instance in terms of directed body movements. The younger Descartes chose to solve this problem by negating it, creating the trenchant dualism of mind and matter. The body-machine would function according to the physical laws of motion, quite independently from the action of the mind.

But why should harmony rule the relation of mind and matter, of body and soul? What was the reason for bodily movements to correspond to mental intentions at every instance, if there is no pathway for any interaction between the two substances? Was the logical structure of Creation enough to explain why the human mind should be able to acquire perfect knowledge of physical laws? Again denying any psychophysical interaction, the occasionalists claimed that the soul was related to its body by miraculous divine intervention. Whenever the occasion arouse that mind appeared to be affected by matter or affecting matter itself, this had to be due to divine intervention coupling cause and effect.

Permitting the continuous intervention of God in His Creation was to admit the action of primary causes, however, and this, in Leibniz's view, threatened the very possibility to do science as much as would the admission of randomness in the search for causality.

Leibniz continued to distinguish extended matter from the unextended mind, and he again denied the possibility of any interaction between these two substances. If mind and matter, body and soul, appear to act in concert, this was not due to a reciprocal influence between the two, but to a *pre-established harmony*, preconceived by God at the moment of Creation. The Creator would have foreseen, harmoniously regulated and preordained all actions of mind and matter for the whole duration of His Creation, checking on all possible combinations of actions and their consequences prior to their actualization and choosing from the many possible worlds the best one. Mind and matter, body and soul: the two substances were conceived of as two clockworks, constructed, wound up and set into harmonious motion by the Creator.

If matter is equated with extension, then it must be infinitely divisible: the mathematical investigations following from this premise were used by Leibniz in his foundation of the principle of continuity which itself can be traced back to Aristotle. The principle of continuity not only served to reject the hypothesis of a vacuum, but also to relate all natural phenomena to an unbroken chain of causality.

Atomistic philosophy had postulated a world constructed of immutable, indivisible and indestructible particles or "atoms". These particles would move through space according to the laws of physics, spontaneously combining to produce temporary organic or inorganic appearances and dissociating again to become freed for new associations. The combination of atoms was thought to be largely determined by the laws of physics or by chance, organization of matter and the phenomenon of life therewith becoming a contingent property of matter. One fundamental claim of atomistic philosophy was the existence of empty space which alone would permit the unconstrained movement of atoms.

If, however, the principle of continuity holds, there can be no empty space in this world. If there is no empty space, all natural phenomena must be embedded within an unbroken chain of causality, leading back to the Creator as to the first and sufficient cause of their existence. If all natural phenomena are causally determined, there is no room neither for miracles, nor for chance. The scientific program thus finds its firm foundation. Were empty space admitted, creating a gap in the chain of causality, the belief in miracles or randomness would get a chance to find their way into scientific argumentation.

The invention of the pneumatic pump by Otto von Guericke around 1649 may seem trivial in retrospect. As a matter of fact it touched on the very foundations of natural science: it served to produce a vacuum on an experimental basis and thus constituted evidence not only against the Cartesian laws of motion, but also against the Leibnizian principle of continuity, both concepts implying the denial of the existence of empty space.

Locke: the Triumph of Empiricism

The philosophy of John Locke (1632-1704) and his followers prepared the grounds for the French Enlightenment. Empiricism was radically opposed to the views of Descartes and Leibniz as it not only admitted a reciprocal influence between mind and matter, soul and body, but even sought the basis of all knowledge in the effects exerted by matter on mind, which is why the original version of empiricism became known as *sensualism*.

Locke and with him empiricists rejected a theory of knowledge based on deductive reasoning, starting out from a set of principles or "distinct", i.e. "innate" ideas. In their view, the human being is born as a *tabula rasa,* and all knowledge results from experience which molds the individual personality: *nihil est in intellectu, quod non antea fuerit in sensu.* This theory is beautifully illustrated by the famous simile of E. Bonnot de Condillac (1715-1780) of an inanimate statue which is gradually brought to life by the succession of sense impressions to which it is exposed. Cognition, based on observation, i.e. on perception, implies the affection of the mind by the observed material objects via the sense organs. In contrast to rationalism, then, empiricism starts out from induction, attempting to formulate a general theory on the basis of a number of particular observations.

Naive versions of (sensualist) empiricism were based on the belief that sense impressions would mediate some image of the material object as it "really" is. The problem is what "reality" can mean in this context. Access to any absolute "reality-in-itself" was effectively denunciated by Immanuel Kant (1724-1804): all we are dealing with are observations, i.e. appearances constrained by the structures of human cognition. Objective reality, as expressed by lawful relations for instance, can relate to appearances only, while the concept of a "reality-in-itself" must remain an incommensurable category of thought.

The difference between rationalism and empiricism implies an even more fundamental disagreement between two schools of thought, concerning the very nature of scientific knowledge itself. Rationalists were striving for certainty of knowledge, for the cognition of absolute truth, for an understanding of the essence of phenomena. This is why they would accept logically stringent deduction only, starting out from self-evident principles and believed to be guaranteed to accurately reflect the properties of the material world by divine intention. Empiricists rejected such a metaphysical bridge to eternal truths. Their program was a science made by man and for man, restricting itself to the hypothetical explanation of perceived phenomena, i.e. to the world as it *appears* to us, rather than attempting to understand what remains inaccessible to human cognition. Empiricists were conscious of the fact that human powers of cognition are not infallible, and they consequently acknowledged that when they evoked natural laws, these were to be understood as a set of hypothetical explanations of observed processes, i.e. of regular appearances. Although they therefore denied the possibility of absolute certainty and knowledge of absolute truth, they still had enough confidence in the regularity of natural processes as to trust in predictions, deduced from theories which themselves are based on a sufficient number of particular observations.

Hume: The Critique of Empiricism

David Hume (1711-1776) sketched the 'problem of induction' in his critique of empiricism, pointing out that the inductive-deductive research program is both unsatisfactory and circular. In the first place there will always remain the problem as to which number of particular observations should be considered sufficient for the inductive formulation of a general theory or natural law. Secondly, empiricism claims to be able to discover by experience the uniformity of natural laws as they determine observable natural processes, but on the other hand the uniformity of natural laws is used as basis for the prediction of processes as determined by these laws (Cassirer, 1973). As Chalmers (1986) put it: The laws of planetary movements might be claimed to have been discovered by induction, i.e. from the observation of these movements, and to have subsequently been used for the successful prediction of solar eclipses. Induction, however, cannot be logically justified by induction, as this amounts to circular reasoning. One way out of the dilemma would be to support a theory of probability rather than a general theory or law by inductive argumentation, but this revision of inductivism still remains subject to the problem of 'induction'.

As induction cannot be justified by logical argumentation nor by the specification of a definite number of particular observations necessary and sufficient for the support of a general theory, Hume turned to scepticism. He concluded that the scientific program cannot be supported on rational grounds, and that the belief in theories or natural laws is nothing but a habit conditioned by repeated experience.

Popper and the Development of Falsificationism

K.R. Popper claimed to have solved the problem of induction "in 1927 or thereabouts" (1972: 1) - with recourse to deduction. Induction does not exist, so he writes; instead, all cognition depends on deduction, all observation must presuppose some theory, however vague this may be. Tell a bunch of students to "observe!", he writes, and they will necessarily have to ask "what?". Disclaiming the call for ultimate truth defended by rationalists, he formulated his falsificationism. According to Popper's philosophy, truth remains a myth. All that science can offer are theories of variable explanatory value. Whether these theories are correct, let alone true, will never be known. What Popper believed to be possible, however, was the test of these theories and their potential falsification. All theories to be accepted as scientific must be testable and at least potentially falsifiable. If the prediction deduced from a theory holds up in experiment, no new information is gained as to the status of the theory in question: verification is impossible. If, however, the theory fails the test, some information is gained after all, since the theory is now known to be wrong: it must either be replaced, or amended. This is how science progresses, by the falsification of cherished theories.

Popper's falsificationism was originally designed in relation to laws of physics. It can be applied in its strict sense only to theories which predict the lawful and uniform regularity of natural processes. These must be reproducible in order to be testable. As a consequence, historical theories, relating to unique processes creating historical contingencies, have no claim for a strictly scientific status under Popper's criterion for the demarcation of natural sciences. Nor are statistical generalizations, i.e. statements on probabilities, stringently testable: they can never be falsified, because they can never stand in logical contradiction to some basic statement. Since statistical generalizations forbid nothing, they can never be shown to be strictly false (Popper, 1976a: 207-208). Sober (1984) has presented a similar argument. In a run of independent tosses of a coin, the probability that the frequency of heads and tails is equal increases with the number of tosses (according to Bernoulli's theorem of large numbers). But: "Frequencies don't *have* to converge on probabilities, even at the limit" (Sober, 1984: 111).

Popper's falsificationism permitted a new criterion of evaluation of alternative hypotheses, replacing Ockham's principle of parsimony. It is no longer maximal economy which characterizes the preferred theory, but its superior explanatory value instead. The explanatory value of a theory is measured by the stringency of a possible test to which it can be subjected, and by the ease of its falsification. Theories protected from test and/or falsification are unscientific.

Two main points of criticism have been raised against the original version of falsificationsim. First of all, the history of science provides many examples demonstrating that a theory was retained although it had been falsified by some particular observation or experiment. Thus, a particular scientific theory may explain some phenomena, but not all available observations. Suppose you have two sets of observations, one revealing a regularity of motion on earth, the other covering the behavior of celestial bodies. A given theory may explain and successfully predict the motion of objects on earth, but it may not provide a satisfactory explanation of all phenomena revealed by the observation of the movements of celestial bodies. Some of these observations may even refute the given scientific theory. On theoretical grounds it would surely be preferable to formulate a new and all embracing theory which accounts for all the available data. For practical reasons, however, the original theory might be retained, because it provides a satisfactory explanation for the phenomena prevalent in the terrestrial environment. A hard time would ensue for many of us - and unnecessarily so - if Euclidean geometry and Newtonian mechanics were dropped in school in favor of the more inclusive relativity theory.

On the other hand, social and political conditions may result in the retention of a scientific theory in spite of its falsification. Lyssenkoism is but one example.There is this famous dictum that theories do not die because of their refutation - rather they die with their last proponents (Klotz, 1980). Popper's research program and his belief in scientific progress appear too optimistic if looked at from a historical perspective (Kuhn, 1973).

The second point of criticism derives from the fact that falsificationism started out from the problem of induction and from the premise that all observation is theory-laden. From this premise follows that observations which seem to falsify a theory under test are themselves theory-laden, and can therefore falsify a theory only to a limited degree. The test of a given theory may depend on an experiment involving more or less sophisticated instrumentalization. This at once introduces a host of new theoretical premises on which the instruments were built and which are supposed to explain and predict their proper function. If there is no assumption-free observation, there cannot be an assumption-free test (Feyerabend, 1981; Chalmers, 1986). This point was forcefully argued by philosophers such as I. Lakatos, T.S. Kuhn and P. Feyerabend, and at a later stage of the evolution of his thinking accepted by Popper himself. Science can thus no longer be viewed as a set of theories tested against a number of "crucial experiments", but must rather be comprehended as a system of competing theories, each with its particular explanatory value but each also with its specific drawbacks. At the bottom of all arguments science boils down to an interplay of competing theories within a given historical, sociological as well as political context.

Empiricism and the History of Science

The study of the history of science shows that scientific progress was neither steady, nor rectilinear. Sensualist empiricism succumbed to the critique of Kant: human perception

14

is constrained by the structures of cognition, and therefore deals with appearances rather than with some hypothesized "thing-in-itself". However, the very notion of "appearance" presupposes something which appears. An up-to-date epistemology reduces empiricism to the hypothetical explanation of regularity of appearances. The historical analysis of scientific progress shows, however, that "what appears" appears differently depending on the way it is looked at: appearances differ with different intentions of the observer working within different cultural contexts.

The positivism characterizing 19th century society has failed: there is no reality which is asymptotically approached by the progress of scientific knowledge. Only that which is perceived can exist, while the history of science as much as modern scientific controversies teach that there are different "ways of seeing" (Hughes and Lambert, 1984). Each of these "ways of seeing" may result in valid explanations of the perceived appearances, but valid only within the constraints of their specific premises. This is why the present treatment of comparative biology as a science stresses the historical perspective, as it alone makes the possibility of different "ways of seeing" clear and obvious. Scientific controversies will not be resolved by an attempt to distinguish right from wrong, since these notions can only relate to the correctness or inadequacy of some theory with respect to its specific premises structuring appearances. Rather it is the analysis of these very premises which matters in the discussion of competing theories.

A summary of the philosophical vocabulary may render the background of competing theories as they are discussed in the ensuing text more intelligible. A first and very basic distinction must be drawn between what is a logical necessity and what results from historical contingency. Procedures of classification may be rooted in logical relations of form, while the theory of phylogeny addresses a historically contingent process. The claim that a logically stringent pattern analysis must correspond to the hypothesized outcome of a historically contingent phylogenetic process is bound to create philosophical problems!

A similarly important distinction is that between premises and observation. Observation reveals nothing but appearances which, if they are regular enough, call for some causal explanation. Philososphical concepts must precede observation, however, if the latter is to transcend mere multiplicity in the search for lawful regularity. If the causal explanation of organic diversity over time and space requires that the mechanisms of evolution as they are observed today were similarly active in the phylogenetic past, this claim must be based on the principle of the uniformity of natural laws, a principle which in itself cannot be based on empiricism. The principle of continuity has a similar status: it must precede observation in order to render appearances intelligible in the context of a continuity of cause and effect.

The precedence of philosophical principles over observation should by itself result in the insight that there is no certainty of knowledge, no access to ultimate truth. All empirical knowledge must remain hypothetical, all science must be understood as a system of competing theories, each with its particular explanatory value, each with its specific drawbacks. The choice among competing theories must depend on some methodological convention. Theories addressing reproducible phenomena may be judged on their predictive power determining the stringency of their testability, as was argued by Popper. Historical theories must be judged on the principle of parsimony.

CHAPTER 2

A REVOLUTION IN BIOLOGY

The publication of Darwin's book *On the Origin of Species by Means of Natural Selection, or the Preservation of Favoured Races in the Struggle for Life* (1859[1959]) is commonly believed to have initiated a major revolution in the science of biology. Darwin figures prominently in the work of historians and philosophers of science, as do Kepler, Galilei and Newton (Thomson, 1986), and justly so. After all, Darwin's book is nothing but a "long argument" to upgrade biology making it meet the high standards set by physics, the paradigmatic science pure and simple. His evolutionary theory makes uniformity of law and continuity of causes compulsory for the study of living beings: "natural selection is daily and hourly scrutinizing throughout the world, every variation, even the slightest; rejecting what is bad, preserving and adding up all that is good; silently and insensibly working, whenever and wherever opportunity offers, at the improvement of each organic being in relation to its organic and inorganic conditions of life" (Darwin, 1859 [1959: 72]). The "law" of natural selection was to have a universal applicability as was generally accorded to Newton's laws of motion and gravity (Schweber, 1985: 38).

However, another and perhaps even more trenchant revolution took place in biology somewhat more than a hundred years before the publication of Darwin's essay on evolution. It will here be argued that it is this revolution which actually opened the way to transformism, and therewith to the Darwinian research program as expressed in the above quotation. Biology was revolutionized following the discovery of *Hydra viridis* by Abraham Trembley in 1740, a discovery which set the stage not only for the reception of materialistic philosophy by students of plant and animal life, but which also influenced views on three basic subjects of comparative biology: the structure of classification, the nature of species, and the mode of reproduction.

Abraham Trembley and His Experiments on *Hydra*

Abraham Trembley (1710-1784), educationist, natural philosopher, and elder cousin of Charles Bonnet, was appointed teacher of Antoine and Jean, sons of count William Bentinck, in 1739. The count lived in a beautiful estate in Zorgvliet near Den Haag, situated close to the seaside. In the summer of 1740, Trembley grasped a handful duckweed from a canal which which traversed the property of the count, keeping the plants in a jar on his windowsill. Demonstrating these plants to his pupils, Antoine and Jean, Trembley observed small organisms, and while he tried to elucidate their nature during the ensuing months, the two boys witnessed a major biological discovery, perhaps the most important one of the century into which they were borne (Baker, 1952).

The organism which Trembley had discovered was *Hydra viridis*. It was an ironic coincidence that Trembley found this species first; had his *polype* not been of green color (due to a symbiosis with green algae which remained unknown to Trembley and his contemporaries), the organism might not have gained the significance it did.

Trembley immediately informed the great René-Antoine Ferchault de Réaumur, author of the multi-volume *Mémoires pour servir à l'Histoire des Insectes*, of his discovery. The two men began a correspondence discussing the question as to whether Trembley's *polype* - as Réaumur called it - was a plant or an animal (Trembley, 1943). The organism was of green color, it was more or less sessile, it reproduced by budding, and its tentacles were first interpreted by Trembley as floating roots of some unknown water plant. In his jar, however, Trembley could observe movements and dislocations of the organism, it proved sensible to irritation, and it was found to catch animal prey with its tentacles. Green color and reproduction by budding was believed to be typical of plants only, while the 18th century physiologist Albrecht von Haller had shown the "*irritabilité*" to be a property typical of animal fibers. To resolve the question, Trembley had recourse to what might be called an "Aristotelian test" of the plant or animal nature of the intriguing "*zoophyte*". Aristotle had argued that plants differ from animals by their ability to regenerate a whole organism from some isolated part. Animals would at best be able to regenerate some lost part which itself, however, would never grow to a new organism again.

Trembley cut the polyps into pieces, finding to his surprise that each of the parts would regenerate a whole new organism. Trembley found no limitation of the number of particles into which the polyp could be dissected without losing its powers of regeneration. He observed that he could graft part of one polyp on to another, and that the organism would reorganize itself if turned inside out. The reports of his experiments to Réaumur, as well as to his cousin Charles Bonnet, sounded so incredible that both men resolved to repeat them, with the same result. Although the Aristotelian criteria of demarcation indicated a plant nature for *Hydra*, Réaumur's authority came down on the side of the animal nature of this "*zoophyte*", mainly by reference to other, marine hydrozoans discovered by Bernard de Jussieu (1699-1777) (Trembley, 1943).

The scientific revolution which Trembley initiated by these experiments resulted not only from the fact that new properties were to be accorded to animal nature; an even more important aspect of his work was that it lent empirical support to the renaissance of atomism which characterized the rise of materialism during the French Enlightenment.

Atomistic philosophy viewed organisms as being composed of parts. Indivisible particles or atoms spontaneously combine to form organized beings, a process which in the view of atomists was not necessarily confined to the female uterus, but which could take place in any suitable environment. Generation, be it spontaneous or sexual, was reduced to the juxtaposition of preformed elements or atoms (Rieppel, 1986a). This view of organization will be discussed in greater detail below, but some fundamental consequences may be mentioned at this juncture already. Firstly, atomistic philosophy is materialistic because organization, and life resulting from it, are contingent properties of matter. Secondly, in the original version of atomism the juxtaposition of parts is guided by physical laws of motion and by chance, and it may therefore be variable or even include 'errors', i.e. result in malformations. The Kentaur of Greek mythology may serve as an appropriate metaphor illustrating the atomistic view of life. Monstrosities may be generated as a result of the fortuitous aggregation of parts. The organization of matter had no inherent (or imposed) *telos*, i.e. life. Rather, vital functions were considered to result from the harmonious combination of parts; should their combination be disharmonious, the organism was doomed to die. As the whole organism is composed of parts, variation is fundamental and can potentially affect the organized being in all its constituent elements. This opens the path to continuous variation, the types of organization grading insensibly into each other. Atomists were therefore ready to accept the concept of a continuous *scala naturae*, although it must be remembered that this concept proved compatible with other philosophical premises, too.

Classification: *l'échelle des êtres*

18th century biology was obsessed by the concept of the Great Chain of Being (Lovejoy, 1936), whose foremost proponent was Charles Bonnet (1720-1793), natural philosopher of Geneva. In his *Traîté d'Insectologie* of 1745, Bonnet gave this rather abstract idea the concrete form of a classification which he sought to improve in his subsequent work, particularly in his *Contemplation de la Nature* of 1764.

As Lovejoy (1936) has shown, the concept of the *scala naturae* is related to the Platonic principle of plenitude as well as to the Aristotelian principle of continuity. In his *Timaios*, Plato dealt with the Creation of the world: man was created first, but during the cycle of his reincarnations, he may degenerate to lower types of organization according to his lack of virtues. Degeneration of man continues through the female to birds, then down to animals living on land, to those living in the sea, etc. According to Plato, Creation thus proceeds downwards along a scale of decreasing levels of organization; Christian tradition reversed the direction, but preserved the same basic idea, i.e. that of an (ascending) ladder of perfection. Since the Platonic demiurgus is perfect, his mind must contain the idea of everything that is possible; if Creation is to mirror the Creator, it must contain the materialization of all his ideas; according to the principle of plenitude, the world must therefore contain everything that is possible.

The Platonic scale of perfection, comprising the actualization of all the Creator's ideas, may be visualized as a line. A line, however, just as time, can infinitely be subdivided, as Aristotle concluded, because points cannot be next to each other (Sorabji, 1983). If the start is made with a given point, a point next to this one cannot be specified, since if this is done, it opens the possibility to hypothesize a point in between which is even closer to the original one. If, however, it is claimed that the next point is no distance away from the starting point, the two points become indistinguishable. From the concept of infinite divisibility results the Aristotelian principle of continuity. Continuity appears to be divisible into discrete units, but this must be an act of abstracting perception, an intended choice of how many units to separate from each other. "Tell me frankly", Simplicio was asked by Salviati in the great *Dialogue* of Galilei, "whether the continuum comprises, in your view, a finite or an infinite number of parts". Simplicio answered that it contains an infinite as well as a finite number of parts, "an infinite number before the continuum is subdivided, a finite number after its subdivision…parts are only *potentially* contained in the continuum". Parts cannot *actually* be contained in the continuum before its subdivision, since the latter cannot be composed of a series of indivisible parts: points cannot be next to each other!

Relating the concept of a *scala naturae* to his principle of continuity, Aristotle had to postulate a gradual transition from inanimate to animate objects. "Nature", he writes in his *De Animalibus Historia*, "progresses so gradually from inanimate things to animate beings, that because of continuity one cannot decide where the boundary between the two divisions should be drawn, and to which of the two sections the intermediate products should belong…the transition from plants to animals is, as already pointed out, a continuous one. In view of many marine organisms one may doubt whether they are animals or plants…". Aristotle cited sessile bivalves and ascidians as examples. The logical consequence of this view of nature was that "…there are animals which, according to their structure, occupy an intermediate position between man and quadrupeds, such as the apes…".

Leibniz, it will be recalled, made the principle of continuity the cornerstone of his scientific program. He used it as a metaphysical foundation, however, claiming that "…among simultaneously existing things continuity may even prevail where sensual perception recognizes nothing but abrupt transitions". The quotation shows that Leibniz asserted continuity to prevail even if such cannot be observed due to the imperfectness of human powers of perception. This plainly shows that the principle of continuity has metaphysical roots and therefore cannot be refuted nor corroborated by observational data. The reason to defend the principle of continuity on metaphysical grounds has moral and ethical dimensions as expressed in a letter to Pierre Varignon, first published by Samuel König on the occasion of his conflict with Maupertuis in 1751: "According to my views everything in the universe is connected in such a way that the present is always pregnant with the future, and every condition can only be explained in a natural way by the immediately preceding state of affairs. If this is denied, gaps have to be admitted to exist in this world which overthrow the great principle of sufficient reason and which force us to have recourse to miracles or to pure chance...". On the principle of continuity, every appearance is embedded in an unbroken *nexus causalis*. The continuous chain of causality ultimately leads back to the Creator as the first cause and sufficient reason for the existence of Creation. Natural science thus becomes related to the moral and ethical codex of Christian tradition. Leibniz furthermore emphasized that, according to his views, continuity prevails not only in the order of succession, but also in relations of the simultaneously existing. Accepting the concept of the *scala naturae* in a metaphysical sense, as portrayed in his *Monadology*, Leibniz believed it to be mirrored in nature, drawing the same conclusions as Aristotle before him: his letter to Varignon contains the noteworthy prediction of the existence of "zoophytes", organized beings intermediate between plants and animals.

The dispute between Samuel König and Maupertuis gained the attention of academics throughout Europe, because Maupertuis was at that time president of the Berlin Academy of Sciences. König used Leibniz's letter to Varignion to demonstrate the philosopher's priority in the formulation of the "principle of minimal action" which Maupertuis claimed to have discovered himself. Through this incidence the letter of Leibniz was widely read, e.g. by Charles Bonnet who cited excerpts of it throughout his works.

Bonnet was an ardent admirer of Leibniz, adopting the latter's ethical standards and therefore the principle of continuity as a foundation for his natural philosophy: "Nature never proceeds by saltations. Everything has its sufficient reason, or its immediate cause. The actual state of an organized body is the consequence of the product of the preceding state, or to put it more correctly, the present state of an organized being is determined by the preceding state" (Bonnet, 1768, vol. I: 4-5). Continuity not only determines relations through time, as during the "*évolution*" (development) of organisms, but also through space, as in the "*échelle des êtres*": "I have gone to great length in my *Contemplations* to discuss this marvellous gradation which reigns between all living beings" (Bonnet, 1769, vol. I: 202). In 1745, Bonnet visualized the Great Chain of Being as an unbroken succession starting with the four Aristotelian elements and progressing through minerals and fossils to plants and animals, culminating in man. In his later writings, he felt that nature seems to leap from minerals to plants (Bonnet, 1764, vol. I: 234), although he acknowledged that this violated the principle of continuity. But, like Leibniz, he had recourse to the imperfect powers of human perception and cognition, cautioning that the apparent gaps may not be real. The transition from plants to animals likewise evoked second thoughts, but here at last he found Leibniz's prediction of the existence of "*zoophytes*" corroborated by the discoveries reported by his cousin Abraham Trembley: *Hydra viridis*, an animal, as Bonnet acknowledged, but with closer affinities to plants than any other (Bonnet, 1764: 199-200). The *polype* served as empirical evidence for the validity of the *universal law* of continuity (Bonnet, 1764, vol. I: 231), which was also confirmed - in the view of the Genevan aristocrat - by the transition from quadrupeds to man: "Humanity shows gradations just as all other productions on our Globe. Between the most perfect human being & the ape there is a large number of continuous intermediates" (Bonnet, 1764, vol. I: 81).

Nature was classified by Bonnet along a continuous linear hierarchy, the Great Chain of Being, ascending to progressively more complex levels of organization. The continuity of this "*échelle des êtres*" had an important consequence, as it implied nominalism. If the products of nature insensibly grade into one another, as they do on the principle of continuity, then all classification boils down to the abstraction of artificial groups which have no actual existence.

"Nature shows no saltations; everything is graded, shaded...among the characters which serve to distinguish the organisms from each other, we find some which are more widely distributed than others. From this follow our classifications in terms of classes, genera and species. These classifications will not truly separate groups, however. There will always be intermediates between two classes or between two neighbouring genera, which will seem to belong to one group as much as to the other. The *polype* relates plants to animals...But, if nothing truly separates in nature, it becomes evident that our classifications are not hers. The classifications which we formulate are purely nominal, & we must look upon them as depending on our needs & on the imperfection of our knowledge" (Bonnet, 1764,vol. I: 28-29).

A few years later he wrote: "Each species has its specific characteristics which distinguish it from any other. The totality of the characters constitutes the *nominal* essence of the species...As the naturalist attempts to classify all organic beings within *classes, genera & species*, he recognizes that the *divisions* of nature are by no means clear-cut..." (Bonnet, 1769, vol. I: 202).

Hydra viridis became the famous "*zoophyte*" of 18th century biology, linking the plant with the animal kingdom and thus corroborating the concept of a continuous "*échelle des êtres*". The *scala naturae* went through various modifications in the writings of subsequent workers, who left it to Lamarck, however, to explicitly separate the inorganic from the organic and the plants from animals. But every author who subscribed to the principle of continuity and therewith to a linear hierarchy reflecting the gradually increasing complexity of organization succumbed to the same logical implication: nominalism. Just as points in a continuum, groups are only potentially contained within a gradual series of forms. All that really exists are individuals. Groups can be delineated and classified, but this must be an intended act of abstracting perception, of ordering intelligence: their definitions are nominal definitions, their characters define names, not "real" things. However, all continuity is conceptualized in terms of a series of discrete steps: the metaphor is that of the "*échelle des êtres*", of a ladder of life, where the gaps between individual rungs must be bridged by the hypothesis of continuity. The principle of continuity is a metaphysical - or methodological - *a priori*, which cannot be refuted by reference to observation of discontinuity. Instead, it is the principle of continuity which guides and constrains observation (Rieppel, 1987a): what are its implications with respect to the nature of species?

The Paradox of the Evolving Species

Darwin's book on evolution was entitled "*On the Origin of Species*": his theory of evolution was designed to provide an explanation of the origin of species, and therewith of the taxic diversity on earth. A closer analysis shows that there was something paradoxical about his "long argument". Asked what a species should be, Darwin (1859 [1959: 469]) answered: "...I look at the term species as one arbitrarily given for the sake of convenience to a set of individuals closely resembling each other...", or: "in short, we shall have to treat species in the same manner as those naturalists treat genera, who admit that genera are merely artificial combinations made for convenience". This obviously is a nominalistic position, as Louis Agassiz made plain in his review of Darwin's essay on evolution: "If species do not exist, how can they vary?" Darwin rebutted in a letter to Asa Gray on August 11, 1860: "I am surprised that Agassiz did not succeed in writing something better. How absurd that logical quibble -'If species do not exist, how can they vary?' As if anyone doubted their temporary existence" (see also Hull, 1973). P.-J.-M. Flourens, however, the successor of Georges Cuvier as Perpetual Secretary to the Paris *Académie des Sciences*, attacked Darwin along similar lines: "Darwin had not offered a definition of species, and yet he claimed it was variable" (Corsi and Weindling, 1985: 701).

The apparent conflict was analyzed by Beatty (1982) who traced Darwin's nominalism back to pre-evolutionary connotations of the term 'species'. Citing Georges Buffon in support of his views, Beatty noted that pre-evolutionary biologists characterized the nature of species by two different aspects: similarity and reproductive coherence. Representatives of a species are similar to each other, an attribute which permits recognition and identification of species through time and space. This similarity may be understood as a consequence of reproductive coherence within a species, which remains reproductively isolated from other entities of its kind. According to Beatty, this understanding of species could not fit the Darwinian theory of evolution, which made it necessary to either redefine the word 'species', or to deny the reality of such entities. Darwin chose the second possibility. (In 1985, Beatty published a sequel to his earlier paper, claiming that Darwin denied the possibility to define the species *category* which was laden with non-evolutionary connotations, but accepted the "reality of recognized taxa" [p. 278] as "chunks of the genealogical nexus", *called* species by his fellow naturalists; see also Ghiselin, 1969.)

Although Beatty's (1982) analysis certainly fits the historical and biological facts, the discussion of the status of species seems to disclose an even more trenchant paradox, or incompatibility of theories, resulting from a clash of essentialistic versus nominalistic 'ways of seeing'. The paradox results from the everyday experience that species appear to constitute entities in the hierarchy of nature which are 'recognizable' or which may be 'discovered', named and related (Bonde, 1977; Patterson, 1982). Discovery, naming and recognition of species must be based on some kind of similarity among its representatives, a similarity which is preserved by the mechanisms of heredity. However, as noted by Beatty (1982: 219): "On this definition, the continued existence of a species necessitated the preservation of its likeness. On this definition, then, a species simply could not evolve while continuing to exist". Characters diagnosing the species become its essence, which by definition cannot evolve (gradually) - since if the essence changes, it no longer diagnoses the same species. As Mayr (1987a: 156) characterized the situation: "If species had...an essence, gradual evolution would be impossible. The fact of their evolution shows that they have no essence".

The alternative view therefore is to permit the (gradual) evolution of species. As was briefly mentioned above and will be more fully discussed below, atomists viewed organisms as being composed of interchangeable parts. Variation thus became fundamental, potentially affecting the organism in all its parts. This opened the path to continuous variation among organisms and therewith to the gradual transformation of species. If a gradual (continuous *sensu* Gingerich, 1984; see also Rieppel, 1987a) change of species is admitted through time or space, the diagnosis of the species dissolves within the continuum of the phylogenetic nexus. A nominalistic view of the species must result. This point may be illustrated by contrasting two competing philosophies of pre-evolutionary taxonomy, namely atomism as opposed to the hypothesis of pre-existing germs.

Charles Bonnet was the foremost proponent of the doctrine of the pre-existence of germs during the 18th century, carrying the hypothesis to ever-increased refinement in the course of his work (Bonnet, 1768; Roger, 1971). The defense of this hypothesis by Bonnet was motivated by theological and political intentions (Marx, 1976; Rieppel, 1985a), but it was also designed to explain the observed stasis of the species-specific form. This in turn was thought, by Bonnet, to result from the developmental and functional correlation of the parts of the organism (Bonnet in Sonntag, 1983: 410, 890). The organism, according to his view, was an integrated whole the development of which could not possibly be explained by the successive juxtaposition of parts, as atomists would have it. The harmonious design of organic structure, effecting "perfect adaptation" (Ospovat, 1981) in a world which bears the stamp of Divine Will and Purpose, would never result from the blind action of physical causes (Bonnet in Sonntag, 1983: 890ff), but had to be the result of Creation *ab initio*. Aurelius Augustinus (1961, vol. I: 91) had already stressed in his exegesis of the *Genesis* that the seed contains in an invisible manner all that is going to form the fully grown tree: the seed provided the paradigm for the temporal actualization of the eternal plan of Creation. The same simile was used (in his *La Recherche de la Vérité*) by P.N. Malebranche, who was considered to be the "great apostle of the pre-existing germs" by Charles Bonnet (Savioz, 1948: 93). Bonnet visualized the germs of all organic beings to have been created at the beginning of this world, their development being reduced to the unfolding ("*évolution*") or actualization of the potential idea of organic form contained within them (Bonnet, 1781, vol. XI: 137).

Of course, Bonnet was aware of the problem of individual variation of organisms, which he explained by a corresponding variation of the germs in his early work (Bonnet, 1768, vol. I: 38-39; see also Rieppel, 1985a). During later periods of his theorizing, however, he shifted to the idea that only the "essential characteristics" of each species were preformed (Bonnet, 1764, vol. I: xxviii-xxix; 1769, vol. I: 185, 327, 362): the doctrine of pre-existent germs thus gained a strong essentialistic background (Rieppel, 1986b).

The essentialistic species concept was bound to create a conflict with Bonnet's adherence to the principle of continuity (Anderson, 1982; Rieppel, 1986b, c). As discussed above, this principle dictates that each individual organism represents a separate rung in the continuous "*echelle des êtres*" (Bonnet, 1764, vol. I: 29). The consequence has been discussed above: species definitions are nominal definitions; species cannot claim reality in any material sense.

In contrast to Bonnet, Georges Buffon was an atomist, performing experiments on spontaneous generation in collaboration with John Turberville Needham, and it was largely against these and other authors such as Maupertuis that Bonnet's polemics against materialism and atheism were directed. Buffon believed to have produced experimental evidence for the spontaneous generation of "*animalcules*" in decoctions prepared with plant and animal matter. This led him to his theory of the "*molécules organiques*" of which the organism would be composed. These organic molecules were believed to partake in an eternal turnover of the biosphere. The dead organic body would decompose into its primary matter, the humus, which provides the nutritive substrate for plant growth. Plants in turn form the basis for the animal food chain. Derived from superfluous nutritive matter, the "*molécules organiques*" would become concentrated in the male and female seminal fluids. These are mixed in the female uterus subsequent to copulation, and a new organism would be formed by juxtaposition of these molecules which are provided by the maternal and paternal bodies. The mixed inheritance of maternal and paternal characteristics thus found an easy explanation as did the origin of malformations by an erroneous juxtaposition of organic molecules. In his theory, Buffon was able not only to account for individual variation, but also for a transformation of organic form and organization as a result of "degeneration", comparable to the formation of graded monstrosities. The gradual "degeneration" of form, most fully discussed in the article on the donkey in Buffon's

Histoire Naturelle" but also implicit in his discussions of the meaning of fossils, of the course of earth history, and of biogeography, was variably attributed to what Buffon called "*espèce*" or "*famille*", the vagueness of his terminology creating problems of interpretation of Buffon as a forerunner of Darwin (Lovejoy, 1959a).

On the one hand, the theory of organic molecules permitted the definition of bisexual animal species by reproductive coherence within but reproductive isolation between species; on the other hand, however, the doctrine of a continuous turnover of these molecules, their juxtaposition to form individually variable offspring, and the possibility of their "degeneration" would have necessarily resulted in a nominalistic species concept, especially if the latter's significance was to cover extended dimensions of time and space. Buffon avoided this consequence with his second criterion of species definition or recognition, that of similarity. Whereas his atomistic model of reproduction explained phenomena of heredity and malformation, it failed to explain the observed similarity of representatives of a given species. Buffon marvelled at the fact that reproduction appeared to perpetually renew a similar form and organization in the endless string of succeeding generations (the Aristotelian view of species, adopted by Buffon in the first chapter of his *Histoire des Animaux*, 1749), and sought for an explanation of this observation. The explanation he offered was his obscure doctrine of the "*moules intérieurs*", conditioning the "*molécules organiques*" derived from superfluous nutritive matter to reproduce the traits of the parental body within which the seminal fluid was being concocted: "and it is by the juxtaposition of these organic particles, derived from all parts of the body of the animal or vegetable, that reproduction is effected, always producing offspring similar to the animal or plant in whose body reproduction takes place, because the juxtaposition of these particles cannot take place unless mediated by the *moule intérieur*...and this is what provides the essence of the unity and continuity of the species" (Buffon, 1855: 566).

With his doctrine of the "*moules intérieurs*", Buffon had introduced an element of preformation (Bowler, 1973) and therewith an essentialistic component into his species concept, preventing the loss of the species category in the continuity of the "*échelle des êtres*". The *moules* placed a developmental constraint on the possible range of variation and degeneration of form, limiting it to the immutable boundaries of a given type. Buffon's solution to the species problem obviously smacked of vitalism. Philosophers such as d'Holbach, who accepted Buffon's and Needham's experiments on spontaneous generation as evidence of atomism, but who were not prepared to soften the resulting materialism by some concept lying beyond the reach of sensualist empiricism, were relegated to pure nominalism. Things are continuously formed, grow, mature and decompose again, the indestructible atoms circulating through the biosphere and providing the substrate for endless revolutions of creations and destructions. Organization is but an ephemeral appearance, resulting from the contingent properties of matter and from the laws of movement, amongst which Newton's concept of gravitation ("*attraction*") figured prominently. There is no essence, no individuality, no stability: the continuity of organization and reorganization of matter provides all intermediates which can be expected in a continuous, finely graded *scala naturae*: minerals, plants and animals differ only by degree, not by their nature. "Oh, human being, will you never understand that you are nothing but an ephemeral success?", exclaimed d'Holbach.

Diderot would go down the same road, at least for some distance (see below), as is beautifully expressed in his *Le rêve de d'Alembert*: "Voltaire may ridicule him as much as he likes", utters the feverish d'Alembert, friend of Diderot and former co-editor of the *Encyclopédie*, "but the *Anguillard* [viz.Needham] was right; I trust my eyes; I see them: how numerous they are! they come and they go!...He [Needham again] saw the history of the world mirrored in a drop of water...everything changes, everything passes, there is only the whole which persists. The world starts over again and terminates on end; each moment it is at its beginning and at its end...". d'Alembert continued in his feverish dream:

23

"...everything is in a state of flux...everything animal is a little human; all mineral is somewhat vegetable; every plant is to some extent an animal. There is nothing precise in nature...and you are speaking of individuals, poor philosophers!...What do you want to say by speaking of individuals? They don't exist, no, they don't exist!...And the species?...The species are nothing but tendencies to the similarity by which they are characterized...To be born, to live and to pass away means nothing but to change form...and what does it matter, one form or the other?".

Darwin, with his theory of evolution based on fundamental variation and natural selection, drew on the same intellectual background: "From echinoderms to Englishmen, all had arisen through the lawful redistribution of living matter...This was rank materialism, and Charles knew it" (More, 1985: 452). Darwin's model of evolutionary change rested on the introduction of variation followed by differential sorting. To explain the origin of variation, Darwin devised his theory of *pangenesis*: parental bodies would produce minute particles or *gemmulae* derived from superabundant nutritive matter and contributing to the formation of offspring. The theory of pangenesis comes close to Buffon's system of "*molécules organiques*", with Robert Brown acting as a possible mediator (Sloan, 1985: 93). The fact that the existence of *gemmulae* has never been positively demonstrated did not, in Darwin's view, weaken his call for empiricism, since chemistry was likewise considered as an undisputably empirical science although it is based on elements which "the eye is unable to see" (Schweber, 1985: 41). Darwin was thus basing his theory of variation on an argument by analogy, the "analogy of nature" invoked by Newton (Hodge, 1985: 229). Finally, it was again the "*zoophytes*" which should provide Darwin with a handle for his adoption of an atomistic perspective. It was his mentor Robert Grant who, in his inaugural lecture at the University College in 1828, drew once more the famous analogy with crystal growth: "...and in the Animal Kingdom the same laws operate in the formation of the silicious crystals [*sic*], which compose the skeleton of many Zoophytes, and the calcareous crystals of many Radiated Animals..." (quoted from Sloan, 1985: 84). If organisms form by the juxtaposition of parts, and if these parts are interchangeable, then fundamental variation of organic structure must follow. The validity of this conclusion was confirmed, in Darwin's view, by his investigation of lower invertebrates: coralline polyps "can change into different forms and patterns of organization...transcend[ing] the limits of several recognized orders" (Sloan, 1985:108). If variation is fundamental, and hence all possible intermediates conceivable, then there cannot exist a true gap between any two related species. Rather, these must be expected to grade into each other. Thence follows Darwin's ambiguity with respect to species definitions, and his denial of the reality of species.

The paradox of the evolving species is that gradual (continuous *sensu* Gingerich, 1984) change implies a nominalistic species concept, while recognition of species implies essential similarity. If species are allowed to change gradually, i.e. through a finely graded series of intermediates (Rieppel, 1987a), in all of their attributes, then their definition and diagnosis can rest on nothing but a subjective subdivision of the continuous genealogical nexus (Gingerich, 1979, 1985). If, however, species are to be recognized or discovered, named and related, their representatives must share some essential similarities which by implication cannot evolve. These essential similarities prevent species from moving continuously through morphospace, imparting constraints on form and organization and thus limiting their appearance to restricted and discontinuous domains of structural stability (Gould, 1983: 362-363). The dilemma persist into modern evolutionary theorizing. Ernst Mayr (1963) is the foremost proponent of what he called the biological species concept, which alone is claimed to fit the Darwinian model of gradual evolution: "Species are groups of interbreeding natural populations that are reproductively isolated from other such groups" (Mayr, 1971: 12). In the course of his work, Mayr has vacillated to specify the biological species concept as "actually" or only "potentially" interbreeding populations, but whichever definition is chosen, the biological species concept remains restricted to one aspect of Buffon's species concept only, viz. reproductive coherence within as opposed to reproductive isolation between species. This creates immediate conflicts if the dimensions of time and space are added. In particular, Mayr (1982: 295) finds no other way to demarcate gradually evolving species from each other than by gaps in the fossil record, "as artificial as this may be".

On its definition by the criterion of reproductive isolation, the biological species concept becomes non-dimensional. The problem of species delimitation arises not only in relation to the time axis, but also in cases of geographically isolated populations: "But what we cannot determine except by somewhat uncertain inferences is whether..." they represent separate species or part of a more widespread polytypic species (Mayr, 1987a: 162). And Mayr (1987b: 219) continued: "...one can only recognize species without any difficulties in the non-dimensional situation. It is only when one deals with isolated populations in space and time that one has to fall back on inferences". One might wonder what these inferences may be, and the answer surely must be: some kind of similarity. And back we are at the paradox of the evolving species: how can Mayr claim that the *fact* of their (gradual) evolution shows that species can have no essence (Mayr, 1987a: 156), when this fact must be inferred from the recognition of these evolving species by some kind of similarity?

Partly to circumvent the problems of the non-dimensional (Rieppel, 1986b) biological species concept, Simpson (1961: 153) proposed the evolutionary species concept. "An evolutionary species is a lineage (an ancestral-descendant sequence of populations) evolving separately from others and with its own unitary evolutionary role and tendencies". Accepting "that old canon in natural history of «Natura non facit saltum»", as Darwin put it, one must again ask the question how one evolutionary species is to be demarcated from its successor. The answer provided by Simpson (1961: 165) proves as unsatisfactory as would be expected from his adherence to the principle of continuity: "Certainly the lineage must be chopped into segments for purposes of classification, and this must be done arbitrarily..., because there is no nonarbitrary way to subdivide a continuous line" (see also Gingerich, 1979, 1985).

A modern proposal (Ghiselin, 1974; Hull, 1976; see Rieppel, 1986b, for a review) to escape Szylla and Charybdis of the nominalist and essentialist 'ways of seeing' is to treat species ontologically as individuals, a solution partly derived from Simpson's concept of the evolutionary species (Wiley, 1981) and to be discussed in greater detail below (section on the ontology of natural entities, chapter 5). Species are considered as individuals by virtue of their spatiotemporal restrictedness and their unique evolutionary role, resulting from reproductive coherence, i.e. from the organization of their parts. Species thus become units of selection and evolution (Hull, 1980). The paradox of the evolving species appears to be solved by this species concept in so far as a species comes into existence (is borne) by a branching speciation event (cladogenesis) and goes extinct (dies) as it undergoes splitting itself, or by terminal extinction. All evolutionary changes occurring within the lineage and between the branching events, reflecting the evolutionary role and tendency of the species (anagenesis or phyletic evolution), has no meaning for species demarcation. In other words, anagenesis or phyletic evolution is by definition ruled out as a mode of speciation (Wiley, 1981; Ax, 1984; Willmann, 1985). Cladogenesis on the other hand appears to provide an objective basis for the spatiotemporal restrictedness of the individuial species. But how can this "objective basis" for species demarcation, i.e. cladogenetic events, be objectified? It must be admitted at this point that the thesis of species *qua* individuals does not solve the paradox of the evolving species either. Species are claimed to be "recognized" or "discovered", but as Heise (1981: 289) emphasized: "...things do not come to us with their lable on them". Consequently, some shared attribute or (essential) similarity is required to recognize the "parts" of an individual species. In other words, synapomorphies or shared derived characters are required to objectify branching events and with them the spatiotemporal restrictedness of the individual. And it is only by these essential similarities that species are recognized "in more than one place at more than one time" (Eldredge, 1979: 16; see also Rieppel, 1986b, and Schoch, 1986). If, however, these shared attributes are allowed to evolve, and if evolution is gradual (continuous *sensu* Gingerich, 1984), then the spatiotemporal restrictedness of species as individuals can no longer be objectified (Rieppel, 1986b).

Models of Ontogeny

The discussion of the species problem has highlighted the conflict created by the assumption of gradual change of identifiable species. Whereas species are said to be discovered or recognized, their gradual (continuous *sensu* Gingerich, 1984) evolution would tend to blur the distinction between two or more species. Individual variability and change as opposed to stasis of form and organization are the result of hereditary and developmental mechanisms. A full understanding of the alternative "ways of seeing" organisms and organization as outlined above requires an understanding of ontogenesis, the mechanisms of which are the third aspect thrown into focus of biological research by Trembley's experiments with *Hydra viridis*. Some comments were already made on the atomistic model of heredity and embryogenesis as opposed to the doctrine of the pre-existence of germs, but a more detailed discussion is necessary to distinguish these theories from epigenesis (Rieppel, 1986a).

The essentialistic background of the doctrine of pre-existence was already mentioned above. In the preface to the sixth volume of his *Histoire des Insectes*, Réaumur (1742: lxvii) reported on the experiments of Trembley with *Hydra*, thereby raising the problem of the divisibility of the soul: "Which kind of soul would this be which, like the body, would lend itself to being cut up into pieces, and which would regenerate itself?" Implicit in this question is the idea that the soul must pervade matter in order to endow it with life. Matter was equated with extension and was thus considered to be infinitely divisible. The soul, however, being immaterial and hence having no extension, could not possibly be subdivided. Trembley's experiments renewed Réaumur's interest in the problem of regeneration, dating back to 1712 when he published an influential paper on the regeneration of amputated extremites of crustaceans. Drawing on the analogy between generation and regeneration, Réaumur had explained the repair of lost limbs by the development of pre-existent "germs of regeneration". Now he proposed a similar model to account for the perplexing regenerative powers of *Hydra*. The problem he was confronted with was more complex, however, since in this case he had to account for the regeneration of a lost soul, too. If *Hydra* was cut horizontally, Réaumur believed that the soul would remain in the head portion where its original seat would be. Regeneration of a complete organism from the head portion therefore posed no problem. The foot portion, however, would be deprived of the original soul, and yet it was observed to regenerate a new head. Réaumur drew the conclusion that the soul, contained within the regenerated head, had to be pre-existent within the "germ of regeneration" along with the soma to which it belonged. This hypothesis was the one adopted by Bonnet (1764, Vol. I: 254-255; 1768, vol. II: 65-70; Savioz, 1948: 92).

Unlike his idol Leibniz, Bonnet was an interactionist, considering the reciprocal interaction between body and soul a necessary condition to account for the phenomena of life (Bonnet, 1760). In what he called a "mixed being" the soul would receive sense-impression from the outer world while itself being able to impart a moving force upon the body. Bonnet was unable to explain by which mechanisms the interactions between the material body and the immaterial soul should be possible, but then again he pointed out that Newton likewise had not been able to explain the mechanism of gravity. He had introduced the concept of gravity to account for observed planetary motions, and so would Bonnet use the concept of the "*être mixte*" to account for the observed phenomena of life. Bonnet adopted sensualism with his claim that the personality of an organic being would be determined by its memory, consisting of accumulated sense impressions. The complexity of the personality depends on the perfection of the organization of the body and its sense organs which in turn determine the perfection of the powers of perception of the soul.

What this view of life boils down to is that the soul constitutes the essence of the organized being. The organization of matter is just an epiphenomenon of life which is tied to the potentialities of the soul. The material appearance of an organic body may change, as it indeed does in the course of its ontogeny, but while this change affects the powers of perception of the soul, it does not touch upon its individual nature: the soul constitutes an element of 'being' the potentialities of which become actualized in a process of 'becoming'. The personality grows more complex with increasingly complex organization of the body, the potentialities of the soul thus becoming more and more fully actualized. The immortal soul as such remains unchanged, however.

Metamorphosis is the appropriate paradigm to ilustrate this essentialistic view of life. As a caterpillar metamorphoses into a butterfly, it radically changes its outward appearance; similarly, every embryo changes its material appearance during the developmental process. This does not imply any change of the essence, however: the metamorphosing, i.e. developing animal preserves its individuality, as it preserves its individual and immortal soul. Admit that the immortal soul of the individual being is pre-existent and encapsulated within the germ from which it develops, admit that the brain is the seat of the soul, and assume that the brain of the butterfly is encapsulated within that of the caterpillar from which it develops, then it becomes apparent that no change of outward, i.e. material appearance can alter "*le moi*" (in Bonnet's terminology) of the organism.

Bonnet used a number of interchangeable notions to designate the developmental process: "*évolution*", "*révolution*", and "*métamorphose*" all meant the same for him. In his *Palingénésie Philosophique*, Bonnet (1769) proceeded to expand his theory of development into a cosmic view of the "evolution" of life on earth. The earth would be subject to a continued succession of global catastrophes or "revolutions" which would destroy all forms of organization on earth. Individual beings were to survive these catastrophes, however, by virtue of indestructible "germs of resurrection", held to be the true seat of the immortal soul and located in the brain. Following each revolution, the germs would "evolve", develop or "metamorphose" into successively higher levels of organization and thus ascend the ladder of life in a predestined manner. By virtue of the ascent to ever-higher levels of complexity of somatic organization, the soul acquires ever-increasing powers of perception and thus attains ever-higher stages of illumination: its potentialities become evermore fully actualized. Yet throughout this process of palingenesis, the individual being would preserve the same immortal soul, i.e. its essence. Bonnet wrote that "if we were able to see a horse, a chicken, or a snake in the form they had at the time of their first appearance in earth history, it would be impossible for us to recognize them" (Bonnet, 1769, vol. I: 258), but this statement does not imply any evolution in the modern sense. Rather, the organisms and the species to which they belong simply changed their outward appearance. As the immortal soul was viewed to constitute the essence of being, Bonnet (1769, vol. I: 253; see also Savioz, 1948: 137) had to conclude: "...each species is a unique Whole, forever persisting; but it is destined to appear, from one period to the next, in the guise of a new form, or of new modifications."

The problem of the soul, and its relation to the body, was one aspect which made the hypothesis of pre-existence necessary on logical grounds - and it highlights the essentialistic background of that doctrine. In addition to metaphysics, Bonnet used genuine scientific arguments to substantiate his model of ontogeny, derived from physiological considerations. His argument aimed at the developmental and particularly the functional correlation of the parts of an organism, what brings him into line with modern holism (Dullemeijer, 1974). The parts of the *"Tout organique"* are "so manifestly linked together and subordinated to one another, that the existence of some presupposes the existence of others" (Bonnet, 1769, vol. I: 355). "The arteries presuppose the veins; both of these presuppose the nerves; and those the brain; which in turn presupposes the heart; and all of these presuppose a multitude of other organs" (Bonnet, 1764, vol. I: 154). "From all this I have drawn a general conclusion which I believe to be philosophical; namely, that the *Touts organiques* have been preformed from the beginning..." (Bonnet, 1769, vol. I: 356).

An important point in Bonnet's theorizing is that he believed to be able to assign predictive power to the principle of the correlation of parts. Commenting on d'Holbach's materialistic, i.e. atomistic *Système de la Nature*, he wrote to his friend Albrecht von Haller on 11 August 1770: "...it suffices to show you a hand and a foot in order for you to guess the whole" (Sonntag, 1983: 890). This statement recalls the methodology of Georges Cuvier, who put the principle of the correlation of parts to effect in his research in paleontology and comparative anatomy some thirty years later. Although Cuvier was not willing to comment favorably on Bonnet's work in his *Histoire des Sciences Naturelles*, except for his doctrine of preexistence, it remains unclear whether Cuvier developed the same principle on his own. His claim to originality may well have resulted from the pressure exerted on a creative young scientist by the scientific community established at the Paris *Muséum* of that time (Outram, 1986). The implications of the developmental and functional correlation of parts and the ensuing holism will be discussed in greater detail in the chapters to follow.

What must be stressed at this juncture is that Bonnet used his argument of the correlation of parts mainly to refute atomistic models of ontogeny, according to which the organism would be formed by the fortuitous and successive juxtaposition of parts, a process governed by the physical laws of movement. In his letter to Haller cited above, Bonnet attacked such materialist views of organization which he deemed atheistic: "The animal obviously forms a Whole all parts of which must have always *coexisted* ... However, movement cannot act except in succession: a cause which cannot but act *successively* cannot produce a Whole which bears the stamp of a Cause acting at once & in an indivisible moment of time" (Sonntag, 1983: 890).

It is remarkable that materialists drew just the opposite conclusions from Trembley's experiments than did Réaumur, Bonnet, and other proponents of the doctrine of preexisting germs. Here was an organism that could be decomposed into its constituent elements, each of which would regenerate a new whole. Parts of one individual could be grafted on to another, so that animals with supernumerary tentacles or with two "heads" could be produced. This provided a model on which to base the study of the inheritance of supernumerary digits in the family of the surgeon Jakob Ruhe by Maupertuis in Berlin (Glass, 1959). Animals were conceived of as being composed of parts, their composition being subject to variation or malformations. Believing in the inheritance of acquired characteristics, as most of his contemporaries, Maupertuis (1753) even went so far as to suggest the multiplication of species by the origin and hereditary fixation of "accidental variations". The occurrence of "accidental variations" was taken as evidence of the fact that organization resulted from the contingent properties of matter and the actions of the laws of movement: the doctrine of pre-existence was thus believed to be refuted (Maupertuis, 1745). The immediate analogy to the generation of organic beings was the growth of crystals: "...mix parts of silver, of nitre and of mercury, and you will wintness the birth of a magnificent plant which chemists call the *Arbre de Diane*, the formation of which differs perhaps from that of a tree only in that it is more exposed. This type of tree relates to other trees such as animals which multiply other than by usual generation, such as the *polypes*..., relate to other animals". The atomistic conception of organization permits fundamental variation and therewith proves to be compatible with the concept of a continuous Chain of Being, the rungs of which differ only by degree, but not in essence. Other materialists converged on similar models: Buffon developed his doctrine of the *"molécules organiques"*, while Needham continued his experiments on spontaneous generation. Denis Diderot, in the *Rêve de d'Alembert*, used his friend Théophile de Bordeu's conception of the structure of glands as a simile for the whole organism: the swarm of bees. LaMettrie, finally, did away with the problem of psycho-physical relations by denying, in his shocking *L'Homme Machine*, the existence of a soul which would correspond to something other than contingent properties of matter. Referring to Diderot, LaMettrie exclaims: "Look at Trembley's polyp for yourself! Does it not contain within itself the causes of its regeneration?" (LaMettrie, 1865: 102). There is no need to have recourse to metaphysical principles such as God. As Diderot put it in his *Lettre sur les Aveugles* of 1749 already: "If nature offers us a knot difficult to disentangle, let us try to avoid cutting it through the hand of a Being which in itself constitutes a problem even less easily resolved".

Here, the sensualist empiricism underlying materialistic philosophy becomes apparent. Leibniz, like his successor Bonnet, strove for an *understanding* of natural phenomena by a comprehension of their essence. Their view of life was strongly finalistic. God had created the material world and its underlying essences. The soul constitutes the essence of the individual being, and it contains *in potentia* all possibilities of development and perfection foreordained by the Creator. Development, be it ontogenetic or palingenetic, served nothing but the purpose of actualization of all these potentialities of the individual: development was considered a process of individuation. Bonnet's conception of organization was holistic, subject to upward *and* downward causation. Not only would nutritive particles add to the mass and function of the organism, but the latter was also determined by the Creator's will from above. God is the necessary and sufficient reason for the existence of organization; an understanding of life was achieved when all its manifestations were traced back to His action through an unbroken chain of causality.

The materialists on the other hand refrained from the attempt to understand the phenomena by comprehension of attributes or qualities which lay beyond the reach of human perception. They were satisfied to provide a causal *explanation* of observable facts: the "How" was given preference over the "Why". Materialists wanted science to be pursued by humans and serving human needs, even if human powers of perception were to remain imperfect and hence all knowledge hypothetical (Cassirer, 1973: 73; Callot, 1965: 73; Roger, 1971: 755-756). The materialists' approach was reductionistic. The problem of organization was reduced to the nature of constituent elements or 'atoms' governed by the elementary laws of physics, the formation of an organized being was explained only by upward causation.

However, the purely materialistic approach did not provide satisfactory answers to all the questions asked by the naturalists of the day, either. If organisms were formed by the variable juxtaposition of particles derived from both parental bodies, phenomena of heredity and monstrosities found an easy explanation. Bonnet had to go to great pains and often rather grotesque theories to explain why the mule could combine the ears of a donkey with the voice of a horse, or why there should be *"monstres par défaut"* and *"monstres par excès"*, which the benevolent Creator could not possibly have foreordained after all (Rieppel, 1985a). On the other hand, the essentialistic background of his doctrine of pre-existing germs was suited to explain the observable constancy of species-specific form: "This term [the pre-existing germ] is usually understood so as to mean an organized body reduced to very small size; such that if it coud be analyzed in this condition, one would observe the same *essential characteristics* which the organized bodies *of the same species* show after their evolution" (Bonnet, 1769, vol. I: 362; italics added). It was the essential characteristics of the species which were pre-existent in the germs and which therefore remained unchanged throughout the cycle of reproduction. Atomists on the other hand found it easy to account for problems of heredity and malformation, but their theory proved insufficient to explain any constancy of species-specific form. This is the reason why most materialists sooner or later fell victim to some kind of vitalism.

Pierre Gassendi, who prepared the ground for the renaissance of atomism with his critique of Cartesian philosophy, combined atomistic philosophy with a vitalism of Aristotelian provenience (Adam, 1955: 161-162; Adelmann, 1966, vol. II: 802). In his effort to explain the repeated reproduction of a given species-specific form through generation, he endowed the atoms with a soul which would remember the atoms' position in the parent body (Adelmann, 1966, vol. II: 776-815). In a similar vein, Maupertuis, following the critique of his *Vénus Physique* (1745) by Réaumur (Roe, 1981: 15), had to endow the atoms, in his *Système de la Nature* (1758), with some kind of "intelligence", "memory", "appetite" and "aversion" to explain the fact that species can be recognized in more than one place and at more than one time. The *"moules intérieurs"* of Georges Buffon, criticized by Bonnet as an "occult" quality, fall into the same line of reasoning. John Turberville Needham, *l'Anguillard*, had recourse to a *"force végétatrice"*, while Diderot evoked his famous *"principe vital"*: the chick, as it emerges from its egg "walks, flies, is confused, escapes, approaches, complains, suffers, loves, desires, enjoys itself; all of your actions, it shows them too. Would you pretend, with Descartes, that it is nothing but an imitating machine? Little children would laugh at you, and the philosophers will answer you that if the chick is a machine, you are one also!" Rejecting the Cartesian analogy between the organism and a machine in the *Entretien entre d'Alembert et Dioderot*, the latter concluded that one has to "imagine in the egg a hidden element", which is some vital principle.

Atomism versus Epigenesis

Bonnet's *Considérations sur les Corps Organisés*, first published in 1762, was largely composed as a critique of the theories of generation put forward by Buffon, Needham and Maupertuis, whom Bonnet rightly characterized as descendants of Epicurus, one of the Ancient founders of atomism. In contrast to his forerunner Democritus, Epicurus had postulated that atoms may spontaneously deviate from their rectilinear motion without the influence of an external force. This introduced the concept of chance (and ultimately of free will) into a world formerly believed to be dominated by necessity. The main concern of Bonnet's polemics was the exclusion of final causes from the interpretation of nature by materialists, and the threat of atheism and vitalism which resulted therefrom. In his arguments, however, the Genevan philosopher and naturalist categorized his enemies as epigenesists, and his correspondence with Albrecht von Haller (Sonntag, 1983) shows that he made no distinction between the enlightened French atomists and such authors as the German naturalist Caspar Friedrich Wolff. This created some confusion of terminology used to denote different models of ontogenesis (Rieppel, 1986a).

Hydra as paradigm for developmental processes exemplifies growth, budding, fission and fusion of parts. All of these processes are elements of a modern understanding of embryogenesis. Atomistic models of generation start out from separate parts, which combine to form the whole. Emphasis is thus placed on the fusion of parts. Development starts out from a heterogeneous primordial matter, i.e. from the mixture of male and female seminal fluids containing the preformed (Roger, 1971) rudiments (atoms) of the organism to develop. If the combination and fusion of parts were an extended process, one element successively being added to the others as exemplified by the growth of crystals, then the model would be incompatible with the developmental and functional correlation of the parts as postulated by Charles Bonnet. It is therefore crucial to note that not all atomists viewed the formation of the embryo as a result of a *successive* combination of its constituent elements. Gassendi had already postulated that the atoms of the two seminal fluids combine instantaneously in order to warrant the developmental and functional correlation of parts.

> "...if the parts of an animal were produced one after another those which were formed first would then impede the fashioning of others because of the very manifold, for most part reciprocal penetration, separation, insertion, and other things of the sort...since the brain or the head, for example, consists of so many veins and arteries it cannot be formed unless there are present at the same time the liver and heart, from which the veins and arteries may proceed right into it..." (Adelmann, 1966, vol. II: 811).

In a similar vein, Buffon (1855: 462) explicitly rejected William Harvey's model of epigenesis (see below) which postulated a successive formation of the constituent elements of the embryo. Although he accepted the analogy with crystal growth (Buffon, 1855: 435), he believed that the formation (as opposed to growth) of the embryo, i.e. the combination of its essential parts ("*parties fondamentales*"), is "a matter of a moment" (Buffon, 1855: 620), for functional reasons. Only more superficial characteristics may be added successively during later stages of the "*développement*" of the fetus (Buffon, 1855: 626).

This is a clear distinction between Buffon's theory on the one hand and the views of Maupertuis on the other, as expounded in his *Vénus Physique* (1749). Acknowledging the influence of Harvey's observations in the seventh chapter of his work, Maupertuis adds the explicit qualification of the *successive* juxtaposition of the preformed (Hoffheimer, 1982: 126) rudiments in the process of embryogenesis. After all, Harvey had observed the *punctum saliens*, the rudiment of the heart, as the first organ to develop in the chick embryo, all other parts being successively added to it during later developmental stages. Although Harvey (1651) did in fact observe a successive formation of the embryo, his model of epigenesis is something quite different from what Maupertuis proposed. Maupertuis' theory of generation, like that of Gassendi and Buffon, is atomistic or, as Roger (1971) characterized it, preformationist (see also Rieppel, 1986a). Preformed rudiments, contained in the paternal and maternal seminal fluids, combine - successively according to Maupertuis - to form the whole: development proceeds from a heterogeneous primordial matter to a heterogeneous adult condition or, in other words, the developmental process is reduced to a process of growth, not a process of differentiation. In this respect, the atomistic models of embryogenesis resembled the doctrine of pre-existence more closely than Harvey's concept of epigenesis, which viewed development as a process of successive *differentiation*, transforming a homogeneous primordium into a heterogeneous organism.

Harvey (1651) undertook the investigation of the development of the chick following the earlier work of Fabrizio d'Acquapendente, whose descriptions he critically evaluated. Harvey's work was based on Aristotelian philosophy and methodology, as he himself acknowledged, but where his observations differed from those of the Great Philosopher, he would not submit to conformity with his master. Aristotle believed the appearance of the heart to precede that of blood. Harvey ascertained the contrary: "But I would not dare to subscribe to the opinion of the Aristotelians who hold that the heart is the author of the blood. For its substance, or parenchym, is born a little after the blood and added to the pulsating vesicles. But I am very much in doubt as to whether the vesicles or *punctum saliens*, or the blood itself be elder...It seems reasonable to suppose, however, that the container is made for the sake of that which it contains, and therefore was formed after it" (Harvey, 1651, transl. by Whitteridge, 1981: 118).

The egg, for Harvey, is a formation *de novo*, a homogeneous primordium successively developing into a complex organism: "These homogeneous parts, I repeat, are not made from heterogeneous or dissimilar elements united together, but they arise by way of generation from similar or homogeneous material, and are differentiated and made dissimilar" (Harvey, 1651, transl. by Whitteridge, 1981: 207). The first substance to appear is the blood: "...out of the colliquament [i.e. the area pellucida] is made the blood, from the blood arises the bulk of the body, which at first looks homogeneous...but in it the parts are traced in outline at first by scarcely perceptible division, and later are divided up and made into organs" (Harvey, 1651, transl. by Whitteridge, 1981: 210). The first organ to appear is the heart, a tiny pulsating spot stained red: "The heart, I repeat, or at least its first rudiment, that is the vesicles and the *punctum saliens*, sets up the rest of the body as a future habitation for itself, and when it is built, it enters and hides itself within it" (Harvey, 1651, transl. by Whitteridge, 1981: 210).

Harvey used the term *epigenesis* to characterize development by successive differentiation:

> "Some [animals] have their parts made one after another, and then, out of the same material they are at the same time nourished, increased and formed, that is to say some of their parts are formed before the other which are formed later, and at the same time they are both increased and formed. Now the construction of these begins from some one part as from its original, and by its help the other members are produced and these we say were made by epigenesis, that is gradually, part after part. And this...is properly called generation" (Harvey, 1651, transl. by Whitteridge, 1981: 202).

Following Harvey, the process of epigenetic development results from two basic morphogenetic mechanisms, viz. budding and subdivision: "...because it is certain that the chick is built by epigenesis, or the addition of parts by *budding out* from one another...the first [part] to exist is the genital particle by virtue of which all the remaining parts do later arise as from their first original...at the same time that part *divides up* and forms all the other parts in their due order..." (Harvey, 1651, transl. by Whitteridge, 1981: 240; emphasis added). The epigenetic process is one of growth and compartmentalization, one organ formed becoming the material cause of the next one to develop. This is a rather different view of generation as compared to the atomistic models detailed above, postulating the simultaneous or successive juxtaposition of preformed rudiments.

The neglect of the distinction between atomism and epigenesis by Charles Bonnet and his allies in their fight against the rise of materialism caused confusion primarily with respect to the interpretation of the work of Caspar Friedrich Wolff, an 18th century epigenesist who was accused of the same errors as Buffon and Maupertuis. Based on his own observations, Wolff (1764) had become convinced of the theory of epigenesis and therefore rejected the doctrine of pre-existing germs. A.v. Haller informed Bonnet on the work of Wolff in a letter dated 18 February 1764. "This Wolf, which I have recommended to you to read about Epigenesis, has just published against you" (Sonntag, 1983: 410). Bonnet replied eight days later. "Whatever may be the critique of Wolf, I forgive him, my dear & illustrious *confrère*...I will not read him; I do not understand German..." (Sonntag, 1983: 410). Had Bonnet read the memoirs of Wolff, he might have recognized the difference between French materialists on the one hand and Wolff on the other. Instead, he had recourse to his repeated argument concerning the correlation of parts: "Truly, if I meditate on the multiple and varied relations which so directly tie together all parts of a *Tout organique*, I am unable to conceive how these could have been formed one another *by apposition*" (Sonntag, 1983: 410). In fact, Wolff (1764: 106) had preempted the criticism of his opponent: "Who is able to imagine that a wall might be produced all at once? Rather, one brick is added to the other, and when the construction of the wall is terminated and the wall has been painted, it looks as if it consisted of a single and continuous piece. This type of construction is considered to be impossible by Mr. Bonnet...". Wolff (1764: 107) rejected the argument that constructional principles of inorganic matter cannot be extrapolated to the formation of living organisms, and specified: "We are not here addressing the question whether a given body is of organic or inorganic nature. Rather we talk of continuity, and the question must be asked whether the principle of continuity...renders it impossible that one part be formed after the other". "How many things in this world look as if they consisted of a single and continuous piece, although they are formed gradually. Look at the icicles hanging from the roof!"(Wolff, 1764: 106).

Indeed, Wolff did not insist on the simile of the wall being constructed by the juxtaposition of bricks, nor did he view organismic growth as resulting from the juxtaposition of preformed rudiments. A true Aristotelian, he accepted the principle of continuity as well as the empirical fact of a successive formation of the embryo and its parts, and he combined the two aspects in his theory of epigenesis: "...such that each part is first of all an effect of the preceding part, and itself becomes the cause of the following part" (Wolff, 1764: 211). Development proceeds from a homogeneous primordium to a heterogeneous condition as a consequence of a continuous succession of cause and effect. To explain embryogenesis, he used the same morphogenetic mechanisms which had already been invoked by William Harvey. He viewed embryogenesis as a process comparable to "vegetation" (Wolff, 1764: 252), specifying: "...the different parts develop one after the other, and they develop in such a way that one part is always secreted from or deposited by the other" (Wolff, 1764: 210). This process is comparable to growth by budding and subsequent subdivision or compartmentalization.

One aspect of Bonnet's critique of epigenesis, lumping the theories of Buffon, Maupertuis and Wolff together, was justified, however, and that was his opposition against the rise of vitalism. If the embryo is a formation *de novo*, and if its parts develop one after and out of the other, epigenesists again had to face the problem how the form, i.e. the type of organization, was preserved in the course of generation. Aristotle considered the immortal soul to be the "principle of knowledge" of the form, immanent in matter (Balss, 1943: 175, 213). Form was the 'element of being' which became actualized during embryogenesis, a process of 'becoming'. For Harvey (1651), the species-specific form was rooted in the eternal world of divine ideas; its preservation during its actualization in the course of embryogenesis was due to divine providence. God was believed to direct the epigenetic process of differentiation using the male and female as instrumental causes for generation (Harvey, 1651, transl. by Whitteridge, 1981: 234), and the soul of the fertilized egg as the guiding principle (Harvey, 1651, transl. by Whitteridge, 1981: 139, 216-217) residing in the blood (Harvey, 1651, transl. by Whitteridge, 1981: 249).

In contrast, Wolff (1764) put the *causa formalis* back into matter where Aristotle originally had it, much to the discontent of Charles Bonnet! From his theory of the development of blood vessels, Wolff (1764) deduced that the circulation of the humors of the early embryo must be directed by a special force inherent in nature. This is his famous *vis essentialis*, directing the flow of the primordial blood through the germ in such a way that the pathways for the later vessels would be carved and hence a matrix produced upon which to build the rest of the embryo. The circulation of body juices is the essence of life, and it must be powered by some force: "...I have called this the essential force" (Wolff, 764: 160). Following the Aristotelian tradition, Wolff considered the possibility that the *vis essentialis* emanates from the heart, which was held to represent the seat of the soul by Aristoteles, but he rejected this hypothesis because the germinal disc of the egg shows signs of structuring due to the carving effect of the blood before the heart is developed (Wolff, 1764: 168).

The Scientific Revolution

Summing up it may be said that Trembley's experiments on *Hydra* had perhaps two most important, indeed revolutionizing effects on comparative biology: the *"zoophyte"* appeared to provide empirical evidence for the principle of continuity on the one hand, and for an atomistic conception of organization on the other. These two aspects could be combined, as was the case in the writings of materialists postulating a continuous *échelle des êtres* resulting from the passing aggregation of atoms into all possible forms of organized bodies.

The principle of continuity and atomistic philosophy revolutionized the species concept. If organisms result from the fortuitous juxtaposition of parts, and if therefore variation is fundamental, potentially affecting organisms in all of their parts, then all possible intermediate types of organization are possible. The supposed reality of species dissolves in a continuous gradation of form - open to a transformist interpretation.

Whether species are discrete entities in the hierarchy of nature or just ephemeral appearances depends on the theory of ontogeny adopted. In 18th century biology three different models were discussed. The theory defended by religious conservatives was the doctrine of the pre-existence of germs, with a strong essentialistic background and a holistic orientation. Progressive materialists upheld atomistic (preformationist *sensu* Roger, 1971) models of ontogeny, with a strong nominalistic background and resulting in reductionism, unless the authors adopted some type of vitalism. The third model was that of epigenesis, proposed by Aristotelians who combined observation with vitalistic explanations. All of these theories of ontogeny have survived into modern times (Rieppel, 1986a), but only atomism and epigenesis will be further considered, since these two models have most profoundly influenced the theoretical background, the methods and thereby the results of comparative biology.

CHAPTER 3

THE RELATION OF HOMOLOGY

Some readers may have been surprised to find that 18th century atomists believed in a female seminal fluid, an idea going back to Hippocrates and at that time still invoked to account for the problem of heredity. It took Albrecht von Haller, an epigenesist converted to the doctrine of pre-existence, to refute that theory (Roe, 1981), although he did not provide a plausible alternative to account for the phenomena of hybridization and malformation either. Only to speak of heredity presupposes some comparison of organisms with one another, permitting to identify traits which are similar and/or dissimilar. Again it is Hippocrates who is "credited with the first systematic attempt at comparative anatomy" in his treatise *On the Joints* (Russell, 1916: 2). A more modern approach is that of Pierre Belon who in 1555 published his *Histoire de la Nature des Oiseaux*. In this work the skeletons of a bird and of man were compared on two opposing pages in a way foreshadowing the concept of homology. Elements of the same relative position within the skeleton, and of similar shape and function, were provided with identical letters in order "to demonstrate how close the affinity is between the two", as Belon put it. This was a remarkable achievement at the time, and the question must be raised whether it was wholly intuitive, or whether there are some underlying assumptions or operational criteria which guided the comparison of the different types of organization. The fundamental concept of comparative biology is that of the relation of homology, as will be discussed in the present chapter.

The Metaphysical Background

In 1790, J. W. Goethe stumbled over a shattered sheep skull lying in the sand of the old Jewish cemetery of Venice, and the idea struck him that the skull might be composed of a number of metamorphosed vertebrae: Goethe's approach to natural history was comparative. Comparative biology seeks to understand the living world by establishing *relations of similarity* between organisms and their constituent parts (organs), relations which are meant to depict order in nature, i.e. a hierarchical distribution of forms which is amenable to rational explanation (Brady, 1985).

During the 18th century, relations of similarity were established with respect to external characteristics of organisms and on the basis of associative thinking. This approach differed markedly from the modern analytical investigation of order in nature, deriving from the methodology developed by Georges Cuvier around the turn from the 18th to the 19th century, and establishing, on the basis of logical subdivision, a subordinated hierarchy of homologies derived from internal structures essential to the functions of the living organism. Although Cuvier's own ideas were alien to the concept of homology, as developed by his friend and later opponent E. Geoffroy Saint-Hilaire, it will be argued below (chapter 5) that his approach to classification implied the concept of homology, too. In retrospect it must be stated that the hierarchical order of nature resulting from this canon of comparative biology cannot be fully understood without appreciation of its metaphysical background, which in fact is rooted in the idea of the *scala naturae* (Marx, 1976).

18th century biologists ordered organic appearances in terms of a finely graded series of forms, associating organisms with each other on the basis of often spurious superficial resemblance. Flying fish provided the intermediate forms between free-swimming fish and birds in the views of both Benoît de Maillet (1749: 316-317) as well as Charles Bonnet: "What the wings are for birds, the fins are for fishes" (Bonnet, 1764, vol. I: 65). Such a proposition seems absurd if viewed from a modern perspective, but it not only reflects the loyal adherence of these authors to the contents of the Book of Genesis as interpreted by Aurelius Augustinus in his *De Genesi ad Litteram Libri Duodecim*, but also their readiness to associate organisms with each other on the basis of superficial but psychologically prevalent similarity. The same point can be made with respect to Bonnet's (1745) claim that the morel provides the intermediate between stones and minerals on the one hand and plants on the other, between the "Etres *bruts* ou *in-organizés*" and the "Etres *organizés & in-animés*" (Bonnet, 1764, vol. I: 21).

Such a classification of organisms, strange as it must appear to the contemporary reader, is nevertheless of great interest from an epistemological point of view, since it documents the dependence of perception and with it of the establishment of relations of similarity on cultural tradition, theoretical premises, and methodological rules. Cultural tradition dictated loyalty to the exegesis of the Bible, methodology permitted the comparison of organisms with respect to readily perceived external characteristics, and the theoretical premise lying at the heart of the concept of the *scala naturae* was the *principle of continuity* derived from Leibnizian metaphysics (Marx, 1976).

The principle of continuity predicted (or rather dictated) the arrangement of organisms in a continuous linear series of forms; should types of organization appear to human perception to be separated by gaps, continuity had nonetheless to be postulated. The principle of continuity is linked to the doctrine of the sufficient reason in Leibniz's metaphysics, as was discussed in the first chapter, and it is this relation which rendered the postulate of continuity so attractive to 18th century biologists such as Charles Bonnet. Only by its embedment in a continuous series of forms, reflecting the continuity of the *nexus causalis*, would each organic form find the sufficient reason for its existence: viewed from the perspective of Natural Theology the unbroken chain of causality ultimately leads back to the Creator as *first* and *sufficient* cause for the existence of the Creation. God enacted the mechanisms, i.e. the guiding principles of ontogeny, which "create" or actualize the organic forms according to His eternal ideas of Creation.

By the arrangement of organisms within a finely graded series of forms, each individual representing a step (un *échelon*) in the ladder of life (in the *échelle des êtres*), clearcut differences between types of organization as well as gaps between species vanished. A corollary of this view of nature is the reduction of the multitude of organic appearances to a unified manifestation of the Creator's will: the *scala naturae* reflects the unity and uniqueness of the Creator himself: "The UNITY of Design reveals the Unity of the Intelligence which has conceived it. The Harmony in the Universe, or the Affinities among the diverse contents of this vast building, prove that its CAUSE is ONE. The effect of this CAUSE is *one* also: the Universe is this effect" (Bonnet, 1764, I: 3). The idealistic background of this body of thought becomes apparent at this early stage of theorizing already. The "Unity of Design" denotes affinities of organized beings which transcend the multiplicity of their material manifestations but are rooted in God's Plan of Creation instead. The principle of continuity permitted the search for essential affinities of forms hidden behind the varied appearances of organized matter.

The relation of the concept of homology to this metaphysical background is established by its derivation from the postulate of a *unité du type* (unity of type), which in turn is rooted in the doctrine of the *scala naturae* (Marx, 1976: 361; 375-376).The idiosyncratic Robinet (1761-1766) derived the continuous Chain of Being from the additive combination of a single prototype according to a preformed pattern, thereby evoking a parallelism of ontogeny and cosmogeny! "All enlightened men, or those of superior faith, affirm that all beings are of the same order, without essential differences between them; that there has never been more than one single being, prototype of all others, these representing all possible variations, multiplications and diversifications of the first", exclaimed Robinet (1766, vol. IV: 2), and specified: "An organ is an elongated hole, a crude cylinder, naturally active: the most complicated organization may be reduced to this simple idea". In a similar vein, Bonnet (1764, vol. I: 242) wrote: "The same general design embraces all parts of the Creation. A globule of light, a molecule of earth, a grain of salt, a mould fungus, a polyp, a mollusc, a bird, a mammal, man are nothing but different expressions of this design which represents all possible modifications of matter on our globe...". The quotation implies the existence of a common structural plan (*bauplan*) which is to be abstracted from the multiplicity of organic appearances on the basis of the principle of continuity, a principle which also underlies E. Geoffroy Saint-Hilaire's concepts of the "*unité de composition*" (Geoffroy Saint-Hilaire, 1825a: 16; 1833a: 66, 89), of the "*fond commun d'organisation*" (Geoffroy Saint-Hilaire, 1833a: 66, 70), or of the "*plan de l'organisation primitif*" (Geoffroy Saint-Hilaire, 1825b: 88). His intention remained the same: the apprehension of the "*unité dans la variété*" of organismic appearances (Geoffroy Saint-Hilaire, 1830: 86-87; 1833a: 69, 75). Should this unifying principle be denied, nature would no longer remain the "*unité philosophique*": "...there would remain nothing but a PLURALITY OF THINGS" (Geoffroy Saint-Hilaire, 1830: 86).

The issue touched on the very foundations of natural science, since it related to the doctrine of the uniformity of natural laws (Geoffroy Saint-Hilaire, 1825a: 16; 1825c: 265), i.e. to the uniform action of secondary causes. E. Geoffroy Saint-Hilaire had interpreted the opercular bone of fishes as a transformation of the auditory bone of mammals, a proposition which was vehemently rejected by Georges Cuvier. Cuvier wrote in his famous treatise on the "*Ossemens Fossiles*" of 1824 (vol. 5, pt. 2, p. 8): "...the auditory bone does not reappear in bony fishes under the form of an opercular bone ... the opercular apparatus is a special characteristic of those species which have received it...". In his reply, E. Geoffroy Saint-Hilaire (1825a: 16) stressed: "Definitively, the opercular bone is not the only element of a vertebrate animal to resist the power of general laws". The only alternative to his hypothesis of transformation of the same primordial matter into a multiplicity of appearances was, in his view, the special Creation of organs: "For it is that which is in fact implicitly declared and proclaimed in the judgements of some naturalists; those, arguing from *a priori*, admit that nature was unable to predispose its primitive plan so as to render its modification in response to altered conditions of respiration impossible; it is therefore put offside, while the solicitude of a second Creation had to imagine another type, adding to those elements known from related families other and new ones particular to the class of fishes" (Geoffroy Saint-Hilaire, 1824a: 423). Geoffroy Saint-Hilaire strove for the establishment of general and lawful relations of form in the field of comparative biology, and this approach is certainly thrown into question if special Creation were admitted, be it as a manifestation of the almightiness of God or of the creative powers of nature, be it a manifestation of the intervention of a first cause or of accidental causes in the natural course of events.

Cuvier's objection to Geoffroy Saint-Hilaire's concept of the "unity of composition" seemed to be well taken, however, since he appealed to empiricism, pointing out that the transformation of any primordial ("*primitif*" in Geoffroy Saint-Hilaire's terms) structure into variously adapted forms or different types of organization is nowhere observed in nature, neither among extant forms, nor in the fossil record. The latter in particular was interpreted by Cuvier to empirically document the absence of intermediates between different types of organization as much as between the species succeeding each other in the course of earth history. Geoffroy Saint-Hilaire had to concede, admitting the fact that the comprehension of the "*unité dans la variété*" of organismic appearances must be based on the *abstracting* powers of human intelligence, abstraction however starting out from observed phenomena - i.e. topological relations of constituent elements. Similarity of organisms is not material, and therefore cannot be revealed by observational literalism, as Cuvier seemed to imply. Rather, the similarity he searched for had to transcend the manifestations of shape (*causa formalis*) and function (*causa finalis*) of the organs compared with each other (Geoffroy Saint-Hilaire, 1830: 121). Geoffroy Saint-Hilaire claimed to have opened new perspectives in biological research by adopting "... une THÉORIE ABSTRAITE et PHILOSOPHIQUE" ("an abstract and philosophical theory") (Geoffroy Saint-Hilaire, 1830: 139). Transformational homologies, termed "*analogies*" by E. Geoffroy, denote an abstract, ideal or *essential* similarity (Geoffroy Saint-Hilaire, 1825a: 14, 15; 1825d: 151): "When we deal with PHILOSOPHICAL DETERMINATIONS, it must be stressed that only one thing can remain invariable, and that is the essence of the element" (Geoffroy Saint-Hilaire, 1824a: 429-430). The essence represents an element of invariance, of 'being', to be abstracted from the ever-changing, i.e. 'becoming', material appearance of the organism. The establishment of transformational homologies, i.e. the comprehension of this essence, is a philosophical endeavor (Geoffroy Saint-Hilaire, 1924b: 258).

The reason for quoting the writings of E. Geoffroy Saint-Hilaire so extensively is to make the point as explicit as possible that his concept of the unity of type is not based on observational literalism claimed to be adopted by Cuvier, but rather results from the abstracting powers of the human intelligence and therefore was judged by him to be a philosophical concept, a relation of similarity transcending sensualist empiricism. It embodied the same metaphysical background as was shown to be implied in the idea of the *scala naturae*. But whereas the principle of continuity provided the metaphysical basis for the establishment of transformational homologies (the *analogies* of E. Geoffroy Saint-Hilaire), their discovery starts out from observed phenomena, observation entailing conceptualization as it rests on the abstracting powers of human cognition guided by methodological or operational rules provided by the "*principe des connexions*", i.e. by the adoption of topographical criteria of similarity. This aspect is emphasized at this point already, because it will later be argued that the approach of E. Geoffroy Saint-Hilaire does not depend on "unbiased observation", but rather constitutes one '*way of seeing*'(Hughes and Lambert, 1984), to which other ways of looking may be opposed, based on a different metaphysical background and guided by different methodological rules.

Pre-Evolutionary Proponents of the Principle of Homology

Homology is a relational concept: organic structure is comprehended in terms of constituent elements and their relations to each other. As such, the principle of homology rests on a decomposition of the developmentally and functionally integrated organism into its constituent elements which are compared with each other on the basis of some methodological or operational rules. Stages are to be demarcated in the course of a continuous developmental process, organs have to be delineated within the complex organismic whole, in order to achieve some basis for comparison. This is well expressed by E. Serres, a former student and later collaborator of E. Geoffroy Saint-Hilaire: "The principal goal of comparative anatomy is to explain the intimate organization of animals; its methods to do so imply a rigorous *determination* of the constituent elements of organic structure. Without such preceding determination of the constituent elements it is impossible to establish the *relations* which constitute the essence of this branch of knowledge" (Serres, 1824: 377). The decomposition of organic structure into its constituent elements and the establishment of relations of similarity among the latter entails an element of cognitive abstraction or of conceptualization which is an attribute of all human perception and therefore cannot disqualify the principle of homology as a scientific concept: "One tends to forget that the knowledge of a single fact is in itself an abstraction; since an object can only be known by the enumeration of its properties, and since these properties can only be appreciated by comparison, the INDIVIDUALITY of a single fact is obviously rooted in the sum of its relations; however, all RELATIONS are ABSTRACTIONS" (Serres, 1827a: 49-50). All observation therefore consists of the establishment of relations (of similarity or dissimilarity), and as such must be intended and guided by some operational criteria. All empirical knowledge must be expressed in some language and therefore must entail conceptualizations on which notions must be based. Homology is not perceived in an unbiased or accidental sense; it must be looked for, and in order to be discovered, methodological rules such as the topological criteria of similarity must be accepted and put to use. This is simply prerequisite in order to transcend mere multiplicity and to discover the unity of type reflecting the unity of the laws dominating the actualization of form during ontogeny (or phylogeny).

While this context is indisputable, the principle of homology still retains enough controversial issues to relativize its function as a unifying principle of comparative biology. First, the decomposition of organic structure into constituent elements entails an element of arbitrariness in view of the continuity of the developmental process and of the functional integration of the organism (Alberch, 1985), and it furthermore may proceed according to different paradigms, viz. atomism or epigenesis. Second, as similarity abstracting from shape and function transcends the material appearance of form, the comparison of the constituent elements must be based on some methodological rules which have to be agreed upon by convention. This is documented by historical analysis which reveals the change of tradition at the Muséum d'Histoire Naturelle in Paris around the beginning of the 19th century, i.e. a conceptual switch from the serial arrangement of organisms to their hierarchical subordination, correlated with the methodological switch from the analysis of external to that of internal characteristics of organization (Daudin, 1926a). But it is also documented by controversial issues of contemporary comparative biology to be dealt with in greater detail below.

The method proposed by E. Geoffroy Saint-Hilaire (1830) and adopted by E. Serres (1827b: 109) was the *"principe des connexions"*, which serves to establish relations of similarity on the basis of topological criteria. A journalist who commented on the debate between Geoffroy Saint-Hilaire and Cuvier in 1830 circumscribed this principle as follows:

> "It is not the separate organs which are similar, but the materials of which these are composed. The materials resemble each other not by their form, nor by their function, but by their number and position, i.e. by the relations they exhibit between one another; in one word, by their connections. The *law of connectivity* permits no caprioles, no exceptions … To form an animal, nature has at her disposal only a limited number of organic elements, of which she may change the shape but not their relative positions. The situation is comparable to a city the map of which has been traced in advance, determining the streets and the number of buildings. The architect is free to infinitely vary the form, size and functions of the buildings, but he cannot change their relative position. This order, this arrangement, these connections are always identical in all animals" (Geoffroy Saint-Hilaire, 1830: 213-214).

Note that the principle of connectivity does not establish material similarity; homology does not imply material identity of constituent elements of organic structures. Instead, Geoffroy Saint-Hilaire was asking for an abstract relation of topological similarity, based on connectivity.

First the paradigms which may guide the comprehension of organic structure in terms of its constituent elements will be discussed; this will be followed by the analysis of the methodological problems involved in the establishment of relations between the latter.

The requirement for decomposition of organic structures into their constituent elements relates the concept of homology to an atomistic background. Indeed, in developing his idea of a unity of type, Etienne Geoffroy Saint-Hilaire seems to have been influenced by recent developments in mineralogy. During his early studies in Paris he met the abbé René-Just Haüy, who had set new standards for mineralogy with his demonstration "that all crystals could be constructed from a combination of a small number of types" (Appel, 1987: 20). Haüy pursued a morphological approach, initiating the search for a similar order in zoology by Geoffroy (Appel, 1987: 90).

E. Serres published extensively on his "*Recherches d'Anatomie Transcendante*", which led him to formulate his "*Lois de l'Organogénie*", organogenesis being based on an atomistic view of organization. In his discussion of organogenesis (Serres, 1827a), he distinguished between two models of growth, viz. juxtaposition and apposition. The distinction was an important one at the time, and had already been discussed by Bonnet (1769, vol. I: 392-410) subsequent to Herissant's studies on the growth of bone and mollusc shells as published in the *Mémoires de l'Académie des Sciences* in 1763. The distinction of different modes of growth was an important issue because it served to differentiate between alternative models of embryogenesis: in analogy to the formation of crystals, atomistic models of ontogenesis predicted growth by juxtaposition of parts, while the doctrine of pre-existent germs assumed the "*développement*" or "*évolution*" of the embryo as a result of intussusceptional growth. Nutritive molecules were thought to be inserted into the net of elementary fibers which would constitute the pre-existent germ, and as a consequence this net of fibers would expand, successively approaching adult size and shape (Bonnet, 1768, vol. I: 56-57). During the later stages of the development of his theory, Bonnet (1769) accepted appositional growth to explain the increase in size and volume of mineralized hard parts which in fact cannot grow by intussusception once the process of mineralization has set in.

In his discussion of the development of the human kidney, E. Serres (1827a: 61) stressed that the organ appears to be a unified whole in the adult: "...the most careful dissection reveals neither sutures nor depressions which might give rise to the idea that it consists of several fractions fused together" (Serres, 1827a: 61). The embryo, however, reveals the existence of eight or ten such fractions which "group themselves according to some laws around the central renal fragment in a manner absolutely equivalent to the growth of crystals. Several ORGANIC INDIVIDUALS fuse to form one single whole" (Serres, 1827a: 61). This model of growth, influenced by Théophile de Bordeu's (1751) views on the structure and development of glands, unquestionably bears the stamp of atomism. For the spinal cord, which shows a stratified structure, Serres (1827a: 61-62) invoked growth by apposition, while the development of teeth demonstrated, according to his views (Serres, 1827a: 62), a combination of appositional and juxtapositional growth. Following his predecessors of the 18th century, Serres (1827a: 62) envisaged the possibility that the universal "laws of attraction", derived from the Newtonian concept of gravity, might serve as a possible mechanism of growth.

The question must be raised whether Serres implied an atomistic view of embryogenesis by his discussion of the modes of growth of organic structures. Such an interpretation of his writing does indeed appear to be admissible. He dismissed the doctrine of pre-existent germs, which he identified as preformationism (for a discussion of these terms see Rieppel, 1986a), on the grounds that organs may change their shape and even their relative position during embryonic development (Serres, 1827a: 57). Although this was not only admitted by A. von Haller, quoted by Serres, but also by Charles Bonnet (1764, vol. I: 292; 1768, vol. I: 122, 253f; vol. II: 252ff), Serres insisted that embryogenesis does not consist of a simple expansion of pre-existent structures, but entails the phenomenon of metamorphosis. The notion of "*évolution*" therefore had to be expanded to cover alterations of form during development, i.e. during *organogenesis*, which means that "*évolution*" had to be equated with "*formation*" (see also Serres, 1827b: 82-83). The reason for the alterations of form during organogenesis is the growth of the organs, which, as was shown above, was interpreted in terms of the paradigm of atomism. This paradigm was therefore also appropriate to explain embryogenesis (see also Serres, 1827b: 121).

The view of the organism as being composed of parts permitted Serres to adopt the concept of homonomy resulting from the metamorphosis of these parts. Whereas homology establishes relations of similarity between constituent elements of different organisms, homonomy (or serial homology) denotes relations of similarity between constituent elements of one and the same organism. Originally (or "essentially") equivalent parts of an organism undergo diverging metamorphosis: "...the different parts of an animal are the repetition of each other" (Serres, 1827a: 52). In support of his views, Serres (1827a: 52-53) quoted Bordeu, who interpreted the symmetry of the right and left sides of the vertebrate body in the light of this concept; Vicq d'Azur, who recognized the homonomy of the anterior and posterior appendages; and A. Duméril, who had the idea to compare the skull with vertebrae: "...the skull was put on a level with vertebrae, just as the anterior limb was put on a level with the posterior one" (Serres, 1872a: 53).

The interpretation of the skull as being composed of a number of metamorphosed vertebrae leads back to the beginning of this chapter, i.e. to the writings of Goethe, who incorporated the concepts of homonomy and homology into an epigenetic perspective. Goethe's theory of the vertebrate skull was based on his *type* concept, first developed in his memoir on the premaxillary bone in quadrupeds and man dating from 1784 (Peyer, 1950). The type, and the concepts of homonomy and homology on which it is based, were not considered by Goethe to result from observational literalism, but rather from a process of observation *and* abstraction. E. Serres (1827a: 49) concurred in the characterization of the methods of comparative anatomy: "Observation is its first tool, abstraction the second". Commenting on the debate of 1830 between E. Geoffroy Saint-Hilaire and Cuvier, Goethe attacked the latter for declaring as arrogance Geoffroy's attempt "to recognize and try to comprehend what cannot be tangibly demonstrated". This statement clearly was meant to curtail the validity of sensualist empiricism, and thus led back to the essentialistic background of natural philosphy which had been abandoned by the 18th century materialists.

Riedl (1983: 208-209) has claimed that Goethe's type concept is really an empirical one, and he corroborated his view with a quotation from Goethe's *Einführung in die Vergleichende Anatomie* (1795) reading: "Experience [sic] must first teach us the parts that are common in all animals [mammals], and how these parts differ. The idea [notion, abstraction] must hold sway over the whole and derive the common picture in a genetic [coherent] fashion. If such a type has been suggested even as a trial [as hypothesis], then we can very well use the usual kinds of comparison to test it" (translation by Riedl, 1983:

208). According to Riedl, Goethe proposed the inductive formulation of a type as it results from the empirical comparison of organisms, and a consequent testing of the type concept by further comparisons. Observation corresponds to the initial data gathering process, whereas the type represents a testable hypothesis emanating from experience. Goethe would therefore have introduced the classical Baconian research program, based on induction and deduction, into the science of comparative anatomy.

This reading of Goethe deserves some comments at this juncture. First of all, there is no experience *per se*; experience must be embedded in some theory (and guided by a method) if it is to be meaningful at all. This is the classical problem of induction, which Popper believed to have solved by the development of falsificationism (see chapter 1). Goethe's natural science was founded on the (metaphysical) principle of continuity which alone could justify the search for unity in the multiplicity of appearances. His approach was therefore embedded in a tradition which guided his perceptions. Even more: he had to use the principle of connectivity to analyze similarity in terms of homology. In other words: the *idea* of unity had to *precede* the empirical search for a common ground plan of organization: the idea of the type provides the standards for anatomical descriptions, wrote Goethe in his *Erster Entwurf einer allgemeinen Einleitung in die vergleichende Anatomie, ausgehend von der Osteologie* (manuscript 1795, first published 1820). Lenoir (1978) shares a similar interpretation of Goethe's type. According to this author, the concept has a regulatory function in the search for unity among the multiplicity of organic appearances. "No individual can serve as a model for the whole" (Goethe, quoted after Lenoir, 1978: 64), nor can the type ever be actualized in nature - as the romantic *Naturphilosophen* would have it. The type, according to Lenoir's (1978) interpretation of Goethe, was an "*Anschauungsform*" serving to transcend mere multiplicity in the aspiration for the formulation of a natural system, however incomplete this classification may be: "Ins Innere der Natur - O! Du Philister! - dringt kein erschaffner Geist" (Goethe, quoted from Lenoir, 1978: 67).

The common ground plan of organization was revealed only if relations of similarity were chosen to be based on topological grounds rather than on form and/or function. The empirical search for the type could not reveal any "raw data", but only appearances, constrained as they are by theoretical and methodological premises. The type concept represents a conceptual construct based in observation, denoting abstract relations of similarity. What is at issue in any test (and potential falsification) of similarity (homology) is not the type concept, but causal or generative mechanisms of form. This however, is no longer Goethe's concern, whose approach to comparative anatomy was "*anschaulich*" (contemplative) rather than causal.

Goethe's thinking first approached the type concept under the influence of the writings of Benedikt de Spinoza. Dealing with the problem of the possibility (or rather impossibility) of interactions between the unextended mind and extended matter, Spinoza claimed that *cogitatio* (mind) and *extensio* (matter) are both attributes of the one and only substance: God (Popper, in Popper and Eccles, 1977: 183). This explains the harmonious relation of mind and matter, of body and soul, by the all-pervading Divinity. As a consequence, Spinoza's philosophy had a strong essentialistic background: reality is attributed not to the ever-changing and hence evanescent material appearances, but to *ideas* which abstract from the contingencies of existence in time and space and treat appearances *sub specie aeternitatis* (Röd, 1978: 189). In a similar vein, the type - an "idea" or "esoteric principle" as Goethe put it - represents an element of invariance, belonging to the category of 'being'; its actualization under a material form corresponds to a process of metamorphosis under the constraints of time, space and causality. The close affinity of Goethe's views with the ideas of E. Geoffroy Saint-Hilaire and E. Serres thus becomes obvious. As Goethe put it in his "*Erster Entwurf einer allgemeinen Einleitung in die vergleichende Anatomie, ausgehend von der Osteologie*", the type represents a *potential* form, which becomes actualized under the influence of internal and external constraints. The internal constraints, according to Goethe, are the *nisus formativus*, a vitalistic concept of a developmental force first postulated by J.F. Blumenbach (1781) and also adopted by Geoffroy Saint-Hilaire (1825c: 265); the *law of compensation*, first developed by C. F. Kielmeyer (1793) and adopted by both, E. Geoffroy Saint-Hilaire (1824c: 146; 1825a: 15; 1833a: 89-90) and E. Serres (1827b: 109); and the "*principe des connexions*": "Dagegen ist das Beständigste der Platz" ("The most conservative is relative position"), wrote Goethe in 1795 already. The external constraints derive from environmental conditions which may influence the actualization of the type in an adaptive sense, as Goethe specified in his discussion of the rodent skeleton figured by d'Alton. The varying actualization of the type is effected by the diverging metamorphosis of originally identical, i.e. equivalent parts of which the organism is believed to be composed: serial homology or homonomy obtains. If the same "typical" or "essentially equivalent" parts metamorphose divergently in different organisms, homology of these parts obtains. The concept of the type evokes two basic issues of 18th century biology: the view of the organism as being composed of parts (atomism versus epigenesis), and metamorphosis.

Homology in Goethe's sense rests on the assumption that an organism is composed of originally identical morphic units, metameres or segments, or, as Starck (1965: 616) put it: the organism represents a *colony* of primary morphic units. As was discussed in the preceding chapter, the paradigm for this view was provided by the discovery of *Hydra* by A. Trembley in 1740. Has Goethe therefore to be interpreted as an atomist? In his treatise "*Vorträge über die ersten drei Kapitel des Entwurfs einer allgemeinen Einleitung in die vergleichende Anatomie, ausgehend von der Osteologie*" (manuscript 1796, first published in 1820), he again compared the formation of organisms to the growth of crystals. Supporting the epigenetic theory of generation as put forward by K.F. Wolff (1764; see chapter 2), and believing in truly creative powers of nature, he noted the difference, that crystals grow by juxtaposition of identical and hence equivalent and interchangeable parts, whereas in living systems the primary morphic units successively emerge one from the other by a process of growth and subdivision (segmentation, compartmentalization). One part forms the material cause of the next one to develop - the hallmark of epigenesis! The astounding regenerative power of plants and lower animals was interpreted by Goethe as evidence of the essential equivalence and loose developmental and functional integration of the constituent elements, permitting an atomistic interpretation of organization. It is only with respect to higher, i.e. more complex levels of organization that Goethe abandoned atomism, a step which in his view was prerequisite for true individuality of animal life. At these higher levels of the ladder of life, the equivalence and thence interchangeability of the constituent elements was abolished by the determination of body axes which polarize organization, and by the functional subordination of the constituent parts as a consequence of their divergent metamorphosis. The result of these advances in organization is that some parts acquire dominance over others, such as the head over the trunk, so that the constituent elements are no longer equivalent and hence interchangeable in the way they are in a crystal.

The notion of *metamorphosis* is the focal point of Goethe's thinking and essential for the proper understanding of his concepts of homonomy and homology. In a letter of 1797, Goethe mentioned having read the work of Jan Swammerdam, which motivated him to pursue personal investigations of metamorphosis in insects. Swammerdam's views on metamorphosis, reiterated in his *Biblia Naturae* published posthumously in 1737, were of central importance for 18th century biology and influenced authors as eminent as Leibniz, René-Antoine Ferchault de Réaumur and Charles Bonnet. Swammerdam viewed ontogenesis as a process of *successive* development (Bowler, 1971) of potentially*preformed* rudiments. He compared metamorphosis with the budding of plants (Swammerdam, 1752: 3, 90), stressing, however, that the animal develops from "invisible yet essential origins present in the maternal body" (Swammerdam, 1752: 115). He further noted that the pupa does not change to become a winged insect imago. Rather it is the same caterpillar, the same pupa, which give rise to a winged animal by the successive development of parts or organs already present in the caterpillar (Swammerdam, 1752: 3-4). Applied to the problem of generation in general, this meant that "there is no generation in nature, but only a development, a growth of parts, and no change is involved" (Swammerdam, 1752: 16). Development or metamorphosis does not imply the origin of something new, but only the actualization of what has always been potentially inherent in the organism. Bonnet's (1760, 1768, 1769) revision of Swammerdam's notion of metamorphosis drastically reveals its essentialistic background: the caterpillar metamorphoses into a butterfly, but in spite of the radical change of its outward material appearance, the animal remains the same *individual*, preserving the same *soul*. The *essence* of the "*personality*" is preserved throughout the whole cycle of metamorphosis. It is in this essentialistic sense that Goethe's notions of homonomy and homology must be understood. The organism is conceived of as being composed of primary identical or equivalent morphic units. Organic complexity and diversity are produced by the divergent metamorphosis of these constituent elements, which results in their individualization and subordination. Yet, in spite of their different form, function, and physiological importance, these constituent elements remain *essentially* the same, i.e. homonomous parts within one organism such as the skull and the vertebrae, or homologous parts between organisms such as the premaxillary bone in man and other mammals.

The goal of this approach to natural history remained the same for Goethe as it was for E. Geoffroy Saint-Hilaire and E. Serres: the organic complexity and diversity of forms is to be reduced to a unity of design which is expressed in the type (Rieppel, 1985b). E. Serres (1830: 7) concurred: "the multiplicity will be reduced to unity by its diverse METAMORPHOSES". Marx (1976: 385) concluded his discussion of Goethe with the remark: "…the biological form is the material and tangible realization of an *idea* … Reducing the multiplicity to unity, recognizing the fundamental archetypes which encompass all manifestations of life … this was the method he chose…".

Methodological Problems of Topographical Homology

The principle of homology depends on the *a priori* acceptance of methodological rules, based on agreement by convention. E. Geoffroy Saint-Hilaire suggested the "*principe des connexions*" as guide to the recognition of homologies. This principle has usually been interpreted to establish operational criteria based on topological relationships of the structures to be compared. Remane (1952), in his classic text on comparative biology, distinguished three principal and three auxiliary operational criteria of homology. The principal criteria are 1) similarity of topographical relationships, 2) similarity of special structure, and 3) continuity of form as revealed by intermediates. Riedl (1975: 69) has convincingly argued, that Remane's (1952) criteria can in fact all be reduced to his first principal one, i.e. to correspondence of topographical relationships within the structural complex (*Lagekriterium*). Special similarity of structure cannot be expressed other than in topological terms, and the recognition of intermediate forms presupposes other, independent operational criteria of homology. As Remane (1952: 39, 43) noted, and as E. Geoffroy Saint-Hilaire had in fact already realized, the concept of topographical homology depends on a correspondence of the number of constitutional elements involved in the comparison, and on the constancy of their connections. One famous example (also figured by Riedl, 1975: Fig. II 17-21) to document the insufficiency of the topographical criterion, and apparently calling for additional operational criteria of homology, is the *descensus* of the testicles out of the peritoneal cavity into a scrotum in advanced mammals. Here are organs, i.e. the testicles, which are considered to be homologous but which change their position relative to the "rest of the body", either during the breeding season only, or permanently. However, their homology is not established on the basis of their gross topographical relation to "the rest of the body", but rather by their connectivity, i.e. on the basis of their relations to blood vessels, spermatic cord etc., and with reference to ontogeny. Another well-known example is the shift in position of the pelvic fins in advanced teleosts: located primarily behind the pectoral fins, they may come to lie in a position in front of the latter. In spite of this change in relative position, the pelvic fins betray their identity, as they preserve their connection to motor nerves derived from posterior body segments (Goodrich, 1930: 138). Thus a point has to be made to distinguish simple topographical relationships as revealed by adult structure from connectivity as revealed by ontogenesis (Fig. 1).

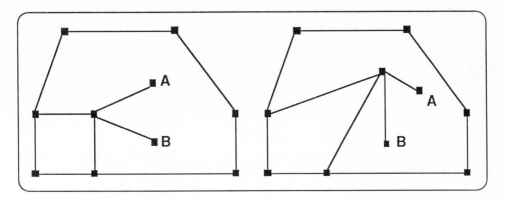

Fig.1: The establishment of homology on the basis of connectivity presupposes the arbitrary fixation of a frame of reference (after Remane, 1952)

The methodological problems of the "*Lagekriterium*" of homology result from the fact that topographic relations must be analyzed with respect to an arbitrary frame of reference. If such is not selected on an *a priori* basis, any number of conjectures of homology would appear possible. This is the reason why Shubin and Alberch (1986) argued for a distinction of connectivity as revealed by the morphogenetic process from topographical relationships as revealed by adult structure. The identification of the metapterygial axis in the tetrapod limb as compared to the sarcopterygian fin has remained so controversial, they note, because "many different axes can be drawn or imagined through the carpal and tarsal elements. The delineation of a particular axis determines the specific homologies that are to be drawn between the fish fin and the tetrapod limb..." (Shubin and Alberch, 1986: 364). The choice of the axis is arbitrary, and it sets the fixed frame of reference in relation to which homologies are determined.

However, the recourse to connectivity as revealed by the morphogenetic process, advocated by Shubin and Alberch (1986), does not seem to resolve these problems either. Investigating the mesenchymal condensations in the developing tetrapod limb, Shubin and Alberch (1986: 333) identified three basic morphogenetic processes responsible for the production of cartilaginous elements: an element may appear *de novo*, in which case its rudiment is not connected to any other. Alternatively, an element may emerge from another either by bifurcation (branching of a single element into two) or by segmentation (a single element gives rise to a distal condensation). The latter two morphogenetic processes establish different patterns of connectivity amenable to empirical observation, which in Shubin and Alberch's (1986: 373) view are more reliable for the determination of homology than topographical relations within adult structural complexes, since "elements with similar patterns of embryonic connectivity can have variable 'positions' within the limb". Closer analysis shows that the concept of connectivity reintroduces the problem of the arbitrary fixation of a frame of reference on a different, i.e. embryonic level. If in urodeles the radius is produced by the humerus through bifurcation, the identification of the radius presupposes that of the humerus as point of reference. Likewise, if the ulnare is segmented from the ulna, how could it be identified if the ulna had not been identified in the first place? And finally, the homology of the humerus throughout tetrapods cannot be established on the basis of embryonic connectivity, since the humerus is described to originate from a *de novo* condensation.

That the choice of the frame of reference within which to establish topographical relations or connectivity does indeed entail some element of arbitrariness is revealed by numerous examples of controversial conjectures of homology documented both by historical analysis and by contemporary literature. One such example derives from the problematical interpretation of caecilian scales. Caecilians (Gymnophiona) are a group of legless amphibians, which initially was classified with snakes. This at least was the contemporary interpretation of Brongniart's (1800) "*Essay d'une classification naturelle des reptiles*". The classification of the Gymnophiona with the snakes was still supported by Cuvier in 1817, although the arrangement originated from an erroneous observation by Cuvier, A. Brongniart and A. Duméril, who jointly dissected snakes and believed their heart to be characterized by the presence of a single auricle only (Daudin, 1926a, vol. I: 184-185).

The problem of caecilian relationships was aggravated by the fact that Oppel (1811) opposed Brongniart's (1800) classification by merging lizards and snakes within one group, the *Squamata*. He defined this new group by its squamous skin, which forms a marked contrast to the "naked" Amphibia, characterized by a soft and moist skin rich in glandular supply, and named *Nuda* by Oppel (1811). As the Gymnophiona show a naked skin, they were included by Oppel (1811) in the Nuda, together with frogs and salamanders. Oppel (1811) supported his classification of the caecilians with reference to a list of characters first enumerated by A. Duméril in 1807, and confirmed by his friend H.de Blainville and himself. As the definition of the Squamata was based on their squamous skin, the apparent nakedness of the caecilians was a readily perceived obstacle to their inclusion within snakes, but Cuvier (1817, vol. II: 86), who referred to them as "naked snakes", continued to support such an arrangement on the basis of the absence of a complex life cycle, involving metamorphosis, in both groups.

However, caecilians do possess scales (Taylor, 1972), as was in fact realized by Cuvier (1817, vol. II: 86), but the problem was to decide whether the scales observed in the Gymnophiona are homologous to those of lizards and snakes or not. Fitzinger (1826), following Oppel (1811), classified the Gymnophiona along with the other amphibians as *Nuda*. He explicitly claimed (Fitzinger, 1826: 4) to have personally confirmed the absence of scales in caecilians, since "by scales we understand regularly formed impressions on the skin, which may be deep or shallow, and which are truly absent in the genus Caecilia ... what Cuvier and Professor Mayer interpret as scales are nothing but rough parts of the skin which are also observed in the genus Siren and its allies" (Fitzinger, 1826: 35). Fitzinger's work was reviewed by Schlegel (1827: 289), who exclaimed: "Mr Fitzinger may understand by scales whatever he wishes, but we must believe that he did not understand what Cuvier or Mayer had said ... it was stated explicitly that the annuli [of the caecilian skin] must be lifted with a scalpel to find a large number of these [scales]". Obviously, the topographical criteria for the identification of "scales" were not the same for Fitzinger as for Cuvier and Schlegel. Fitzinger identified scales as superficial structures, what indeed they are in the Squamata, and he was therefore correct with his claim that comparable scales are lacking in the Gymnophiona. Judging topographical relations on different grounds, Cuvier, Mayer and Schlegel found no reason, however, not to classify caecilians with snakes only because the scales of the two groups have a different position, those of the caecilians being embedded within the skin rather than lying superficial to it. Today, the scales of the Squamata are identified as epidermal structures, those of the caecilians as dermal structures, and they are therefore considered to be non-homologous. The historical analysis shows, however, that this judgment is based on a conventionally defined frame of reference, taking the ontogenetic development into account in addition to topographical relationships (see below for further comments), and that other definitions are possible. (This point will be raised again in the discussion of congruence as a possible test of homology.)

The same point can be illustrated by the polemics between W.O. Kowalevsky and K.E. von Baer which took place after the publication of Darwin's essay on evolution in 1859. Kowalevski had investigated the embryology of ascidians, and his conclusion was that vertebrates might be descended from ascidian larvae. This claim was based on the conjecture of homology of the spinal cord in both groups; the larval spinal cord of ascidians was compared to that of vertebrates on the basis of its topographical relation to an underyling structure which Kowalevski identified as a notochord. Kowalevski's hypothesis of descent prompted the reaction of von Baer (1873) who at the end of his life remained a staunch opponent of Darwinism. His opposition was directed not so much against a hypothesis of transformation, but rather against the materialistic outlook inherent in the Darwinian theory of evolution. Von Baer ascertained that Kowalevski's claim can only be supported if dissimilar things are made similar by the conjecture of homology. Indeed, Kowalevski had himself pointed out histological differences in the structures which were compared. But this, in fact, was not the central issue of the disagreement. The main problem was that the two opponents adopted a different frame of reference for the establishment of homology. Von Baer converged on Cuvier's classification which grouped the ascidians within the *"embranchement"* of the molluscs. This, however, implied that ascidian larvae could not be orientated the way Kowalevsky had it. In von Baer's view, the nervous system of larval ascidians was in a morphologically ventral (if functionally dorsal) position, and therefore could not be homologous to the dorsal spinal cord of vertebrates.

Kowalevsky and von Baer selected a different morphological orientation in their comparison of ascidian larvae with vertebrates. The choice of the particular orientation resulted from the acceptance of a particular hypothesis of grouping, the ascidians being related either to vertebrates (Kowalevsky), or to molluscs (von Baer). The example demonstrates not only the conventional or arbitrary choice of the frame of reference within which homology must be analyzed, but it illustrates furthermore that the acceptance of the frame of reference may depend on a conjecture of grouping based on other characters. In other words, homology is no inductive clue to phylogenetic relationships; instead, it must be established deductively, i.e. in the light of a hypothesis of grouping (Rieppel, 1980; 1984a), a point which will be discussed in more detail in relation to the demarcation of homology from convergence.

The last example to be discussed in the present context illustrates the possibility of the change of topographical relationships of given structural elements. (It has been argued above that a point must be made to distinguish topographical relations from connectivity [Shubin and Alberch, 1986]. The brief discussion of the embryogenesis of the tetrapod limb, abstracted from Shubin and Alberch's [1986] account, demonstrated, however, that connectivity must likewise be related to some fixed point of reference, if an infinite regress is to be avoided.) One such case is the epipterygoid bone in the amphisbaenian genus *Trogonophis*.

The epipterygoid bone is a characteristic feature of the lizard skull: it has the shape of a slender column (called *columella cranii* in the older literature), rising up from the palate (pterygoid bone) to the skull roof (parietal). Embryologically, it derives from the splanchnocranium (mandibular branchial arch), as it represents the ossified ascending process of the palatoquadrate. The bone is an important landmark in the tetrapod skull, defining the lateral wall of the *cavum epiptericum* as identified by Gaupp (1900), an extracranial space, housing the Gasserian ganglion of the trigeminal nerve, and traversed by the deep (ophthalmic) branch of this nerve as well as by the head vein. The maxillary and mandibular branches of the trigeminal nerve pass out of the cavum epiptericum behind the epipterygoid bone and then pass lateral to it as they turn anteriorly to supply the upper and lower jaws with sensible nerve fibers. In addition to its derivation from the splanchnocranium, the homology of the epipterygoid appears to be easily established on the basis of its topographical relations to the branches of the trigeminal nerve, as the bone lies lateral to the deep (ophthalmic) branch but deep to the maxillary and mandibular branches. The homology of the epipterygoid bone is an important issue in comparative anatomy, since it is prerequisite for the understanding of the closure of the lateral braincase wall in birds as opposed to mammals - a problem which will be raised again in the discussion of character congruence as a possible test of homology.

Amphisbaenians are a distinct group of legless squamate reptiles of as yet unspecified relationships to either lizards or snakes (Gans, 1978). Their skull is highly modified in adaptation to their burrowing habits (Gans, 1960). One consequence of these modifications is the closure of the lateral braincase wall. The epipterygoid bone is absent in all amphisbaenians with the exception of the genus *Trogonophis*, where a reduced *columella cranii* has been described (Bellairs, 1949). However, the epipterygoid of *Trogonophis* shows different relations to the branches of the trigeminal nerve as compared to lizards: the maxillary branch passes medial to it (Bellairs and Kamal, 1981: 130-131). Topographical relations are thus demonstrated to be liable to change. One might argue that the nature of the epipterygoid bone of *Trogonophis* might still be revealed by its derivation from the splanchnocranium, if developmental stages were to become available, but this is not the crucial point in the present context. The point is to show that topographical relations of the epipterygoid can change, and that this can become an important problem when it comes to the interpretation of the lateral braincase wall in birds as opposed to mammals (see below). Furthermore, it must be stressed that in order to make such a statement, a fixed frame of reference must already be chosen. To say that the epipterygoid changed its relations to the branches of the trigeminal nerve in *Trogonophis* implies the choice of the maxillary branch as a fixed point of reference, which in turn presupposes the identification of the nerve with respect to some other frame of reference. Alternatively, it might be claimed that the nerve changed its course, but such can only be postulated if the epipterygoid bone is held in a constant position with respect to some other structures. And if the frame of reference for the establishment of its identity is its derivation from the palatoquadrate, the question must be asked how the mandibular arch is identified as such in the first place. Even an experimental approach to homology would not help. In view of the fact that the splanchnocranium is derived from neural crest cells, an experimental approach to its

48

development might appear feasible. However, such an argument not only involves the assumption of a constancy of developmental pathways and tissue interactions, but it also presupposes the identification of the neural crest, which must again be based on topographical criteria.

The arbitrary choice of a fixed frame of reference is only one source of confusion and controversy in the application of the principle of homology; the other is that the methodological rules necessary to establish relations of similarity must be accepted by convention. Authors dissatisfied with the ambiguity of the criterion of topographical relationships might turn to developmental processes as a basis for the establishment of homology. This again is rendered controversial by the fact that there are many examples demonstrating not only the change of developmental pathways of apparently homologous structures, such as the formation of the neural tube by delamination in teleosts and by invagination in amphibians (Alberch, 1985, deQueiroz, 1985), but also the change of inductive tissue interactions which produce homologous structures. The formation of the lens of the vertebrate eye is usually induced by the optic cup approaching the epidermis; however, in *Rana esculenta*, the experimental removal of the optic cup does not interfere with the normal differentiation of the lens, the inductive impulse for its formation emanating from the heart mesoderm (Raff and Kaufman, 1983: 148; Alberch, 1985: 49).

The Interpretation of Homology

The development of the concept of homology as a clue to the *"unité du type"* in pre-evolutionary comparative biology was extensively discussed above. Quotations from the writings of E. Geoffroy Saint-Hilaire, E. Serres and J. W. Goethe have made it clear, so it is hoped, that pre-evolutionary biology borrowed from the Platonic world of ideas, searching for invariable relations of similarity behind the different appearances of organic structure. Relations of similarities were interpreted as the expression of "natural affinities" of organisms, but these affinities were not understood in a material sense, i.e. as the expression of historical continuity of form caused by descent with modifications. If descent with modification was at all accepted, as by E. Geoffroy Saint-Hilaire and his collegues, then only within the limits of a strictly circumscribed *type* of organization (Geoffroy Saint-Hilaire, 1828a; 1833a: 63, 67). The type set insurmountable restrictions to variability, but within the constraints of the type the constituent elements of organic structure were admitted to be subject to adaptive modifications under the influence of environmental conditions acting on the process of organogenesis (E. Geoffroy Saint-Hilaire, 1825b: 89; 1825e: 241; 1830: 11). And it is the principle of homology which permitted the disclosure of the *"unité dans la variété"* among the multiple material actualizations of the type. The "affinities" of organisms manifest in the unity of type were not considered to be rooted in the material appearance of organisms. Comparative anatomy was "transcendental", as E. Serres put it, and the "affinities" which were the object of its investigation were interpreted to express a shared essence of form.

In more modern terms, it might be said that the relation of homology, based on the *principe des connexions*, denotes *logical* relations of form on a rather abstract level. Similarity is based on topological relations (abstracting from shape and function) of constituent elements (abstracted from the developmentally and functionally integrated whole). "Homology is an equivalence relation of a set of forms which share a common structural plan and are thus transformable one into the other. This is therefore a logical relation, independent of any historical or genealogical relationships which the actual structures may have" (Goodwin, 1984a: 101). The reason why the type is a philosophical concept, representing an element of 'being' in the ever-changing or 'becoming' multiplicity of organic appearances, is because it is based on logical relations rather than on historical and hence material contingency. These logical relations, however, may become the object of causal explanation, in particular if they are taken to represent law-like relations of the materialized form. One possible explanation for the apparently law-like relation of similarity of form is provided by the assumption of the creative action of a First Cause, as was common in pre-Darwinian biology.

The order in nature disclosed by the relation of homology lent itself to the incorporation into Natural Theology as manifestation of the Creator's plan, or into the grandiose scenarios of the *Naturphilosophen* (Russell, 1916; Farber, 1976: 108-109). Where E. Geoffroy Saint-Hilaire and E. Serres were searching for "transcendental" laws of form determining relations of similarity of organic structures, others found evidence for the creative power of a transcendental being. As the divine ideas are an attribute of the eternal Creator, they are immutable, representing a category of 'being' - to which the essence of form therefore also belongs. The type, the essence of form, is immutable, i.e. not subject to the constraints of time and space as are its varied material manifestations, which is the reason why a unity of type can at all be abstracted from the multiplicity of organic appearances. The latter in contrast are always changing, i.e. belonging to the category of 'becoming', being subject to birth, growth and death and to the influence of environmental conditions affecting the process of ontogenesis. The cycle of reproduction, duplicating organic form within the fixed limits of the type, permits the partaking of ephemeral and variable individual organisms in the eternity of divine ideas. This combination of Platonic and Aristotelian elements of thought tinged with Christian dogma characteristic of Natural Theology persisted well into the post-Darwinian area under the defense of Louis Agassiz (1807-1873).

Agassiz, like von Baer before him, was opposed to the materialistic background of Darwin's theory of evolution, and therefore argued against this theory on the basis of various grounds, particularly paleontology and zoogeography, thereby defending a finalistic outlook which alone promised the discovery, in the book of nature, the handwriting of the Almighty (Agassiz, 1859: 67). His opposition against Darwinism did not weaken until late in his life, when "he had learned to play the role of the devil's advocate with a challenging skill that could contribute real insights and encourage deeper analyses by those who accepted Darwin's formulations" (Lurie, 1960: 373).

In Agassiz's view, the fossil record documented the fulfillment of God's will, and thus served as empirical evidence for His existence. The rocks revealed the actualization or materialization of the plan of Creation in the course of Earth History. Agassiz viewed the materialization of God's plan of Creation as an ongoing process of 'becoming', postulating a *creatio in principio* (Rudwick, 1972). The evidence for this was not only derived from studies of ontogeny, but also from the study of paleontology. His adherence to Augustinian neoplatonism is perhaps best illustrated by his claim of the existence of so-called "*prophetic types*" among fossils. This concept was first introduced on the occasion of the meeting of the 'American Association for the Advancement of Science' in 1849, in an address which was on that occasion already judged to be provocative by the audience (Lurie, 1960: 162). Embedded in a grand synthetic view of a harmonious nature, which exhibits a parallel "development", i.e. a parallel ascent to ever higher levels of complexity in anatomy, paleontology and ontogeny (Agassiz, 1859: 198; Lurie, 1960: 84), the "*prophetic type*" designated a type of organization documented by some fossil at an early time of earth history which reveals God's intentions by foreshadowing the increase of complexity which would be attained during later geological epochs:

"We have seen ... how the embryonic conditions of higher representatives of certain types, called into existence at a later time, are typified, as it were, in representatives of the same types which have existed at an earlier period. These relations, now they are satisfactorily known, may also be considered as exemplifying, as it were, in the diversity of animals of an earlier period, the pattern upon which the phases of the development of other animals of a later period were to be established. They now appear like a prophecy in those earlier times of an order of things not possible with the earlier combinations then prevailing in the animal kingdom, but exhibiting in a later period, in a striking manner, the antecedent consideration of every step in the gradation of animals" (Agassiz, 1859: 175-176).

It is in this sense, Agassiz explains, that flying reptiles (pterosaurs) announced the rise of birds, and ichthyosaurs announced the rise of dolphins at an early stage of earth history. A leading paleontologist of his time, Agassiz could not deny the fact that organization changed over time. This change was interpreted by him as a general increase of complexity which culminates in the last organism to appear on the surface of the globe, i.e. in man himself, the ultimate term of earthly perfection (Agassiz, 1845: 7). The progressive development of organized beings could not, in Agassiz's view, be reduced to the "blind action" of physical causes, as Darwinists would have it (Agassiz, 1844: xxi; 1859: 10, 15; Lurie, 1960: 152); rather, it had to be intended by God, whose will becomes progressively actualized during earth history according to His providence. And fossils document that it pleased God to announce future increase of complexity at an earlier time of earth history by the actualization of "prophetic types".

Post-evolutionary comparative biology retained the pattern of order in nature elaborated by the "transcendental" morphologists (Russell, 1916; Patterson, 1977; 1981a; 1981b), but the *interpretation* of homology changed with the reception of Darwin's theory of evolution. Homology was no longer thought to be the expression of hidden bonds, but now was interpreted as material evidence of evolution: the First Cause was replaced by secondary causes, logical relations of form became evidence of historical relations by descent with modification. Darwin (1859 [1959: 350]) thought that by his theory of evolution, "...the grand fact in natural history of subordination of group under group, which from its familiarity does not always sufficiently strike us is in my judgement fully explained". As Brady (1985) has cogently argued, the subordinated order in nature is based on relations of homology, and it was so familiar to Darwin's contemporaries because it has long been known by the naturalists of those days, being based on pre-evolutionary comparative anatomy. Brady (1985) justly insists on the point that Darwin never claimed the subordinated hierarchy of types to be an outcome of his theory, but always was clear about his achievement to have provided a causal *explanation* for a pattern which had been observed long before him. Darwin and his successors *explained* the unity of type by unity of descent (Brady, 1985; Shubin and Alberch, 1986).

"Naturalists try to arrange the species, genera, and families in each class, on what is called the Natural System. But what is meant by this system? Some authors look at it merely as a scheme for arranging together those living objects which are most alike, and for separating those which are most unlike; or as an artificial means of enunciating, as briefly as possible, general propositions ... But many naturalists think that something more is meant by the Natural System; they believe that it reveals the plan of the Creator; but unless it is specified whether order in time or space, or what else is meant by the Plan of the Creator, it seems to me that nothing is added to our knowledge ... I believe that something more is included; and that propinquity of descent - the only known cause of similarity of organic beings - is the bond, hidden as it is by various degrees of modification, which is partially revealed to us by our classifications" (Darwin, 1859 [1959: 351]; see also Brady, 1985: 119).

This passage from the *Origin of Species* illustrates that Darwin had no objections to the methods of pre-evolutionary comparative anatomists, nor did he question their results; but he differed from the "transcendental school" by his interpretation of those "hidden bonds" revealed by homology, believing in a causal explanation by propinquity of descent.

On Darwin's interpretation, homology becomes the material expression of the historical continuity of form on the basis of descent with modification. A category of 'being' in the context of idealistic morphology, it now changes into a category of 'becoming', subject to the contingencies of time and space. In spite of this change of interpretation, the *intention* for the establishment of relations of similarity remained the same, however, i.e. the disclosure of the "*unité dans la variété*". In contrast to Natural Theology, which comprehended the multifaceted Creation as an emanation from the one and only God, Darwinism seeks to reduce the diversity of organic life to a unity of descent: evolutionism has preserved the pretension to comprehend the world of multiple material appearances as a unified whole. In spite of its turn to materialism, evolutionism has thus not succeeded to overcome the metaphysical background which was already implicit in the idea of the 'Great Chain of Being' (Lovejoy, 1936; Marx, 1976).

The reason for this extensive discussion of the concept of homology in a historical perspective is that the latter renders the dual meaning of this concept, as a category of 'being' and of 'becoming', intelligible. The multitude of organic appearances is reduced to a common structural plan - subject to changing interpretations over time - by the abstraction of relations of similarity on the basis of methodological rules. These specify topographical criteria or connectivity. Connectivity in turn must be rooted in topographical relations, if an infinite regress in the analysis is to be avoided. The methodological rules therefore specify homology in a topological sense, establishing relations of coexistence (Rieppel, 1985b). Homology in a topographical sense, specifying the hierarchical order in nature, is equivalent in pre- and post-Darwinian biology, which it can be because it is a category of 'being', a problem of *Anschauung* amenable to different explanations.

By their explanation on the hypothesis of evolution, topological correspondences become evidence of common descent. The logical and hence atemporal relations of coexistence change to temporal and hence contingent relations of succession: topographical relations of similarity become homologies defined on phylogenetic grounds.

"*Homologous* features (or states of the features) in two or more organisms are those that can be traced back to the same feature (or state) in the common ancestor of these organisms", is the definition adopted by Mayr (1969: 85). It has frequently been argued that this widespread definition of homology, giving the term its "factually and logically widest possible meaning in the context of the theory of evolution" (Ax, 1984: 167), is circular, because homologous features are to provide the evidence of common descent in the first place. The usual reply to this critique is the requirement to distinguish the *definition* of homology from the *operational criteria* on which particular conjectures of homology are based. This discussion implies a confusion of the *explanandum* (the thing to be explained) and the *explanans* (the explanation), as Brady (1985) has shown: what is observed are topological relations. They represent the "thing to be explained" by a theory of evolution which in turn provides the explanation. This dual meaning of the notion of homology does justice to this relation to observation and interpretation (explanation). What is observed, i.e. conceptualized on the principle of connectivity, are topographical relations of similarity - as opposed to non-homology. By their interpretation as evidence of common ancestry, topological correspondences become homologies in a phylogenetic sense - as opposed to convergent similarity (Rieppel, 1980).

The Congruence of Characters

Like all scientific endeavor, the analysis of order in nature must be based on theoretical premises and methodological rules: if the problems of comparative biology could be reduced to this statement, little argument about order in nature would be expected. Agreement on theory and methods within the scientific community should promote the gradual accomplishment of the goal of comparative biology, i.e. the asymtotical approximation to the "*système universel des êtres*" which Linné was already striving for (Callot, 1965: 403). However, historical analysis shows that this is an overly simplified view of the structure of scientific progress. Not only is the acceptance of theoretical premises and methodological rules dependent on historical, sociological, and psychological parameters, but also are the observational data problematical in themselves.

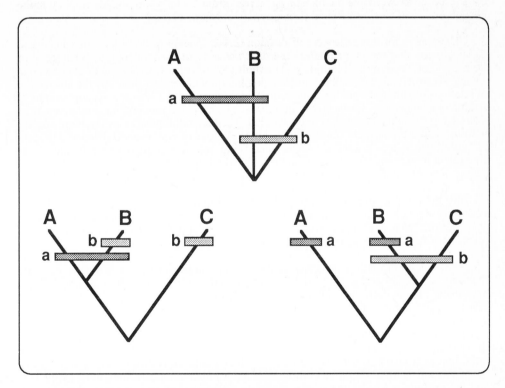

Fig. 2: Incongruently distributed characters a and b support conflicting hypotheses of grouping of the taxa A, B, and C.

All observation consists in the establishment of relations and thus entails an element of conceptualization. As observation must be intended, it must be guided by some theory or expectation in order to be meaningful at all. Which is the intention, the initial hypothesis in the analysis of order in nature? One possibility is to search for a hierarchical representation of the observational data (Brady, 1983), since this allows their most economical representation. This initial hypothesis would be corroborated if characters were regularly falling into a congruent hierarchical pattern. Accepting evolution as a scientific theory explaining order in nature, the question must be asked which predictions this theory might be capable to generate. The basic prediction surely must be that the unique historical process of descent with modification would yield a unique pattern of order in nature (Riedl, 1975). Organisms would have to fall into an uncontradicted pattern of groups within groups.

Such, however, is not the result of observation. A thorough analysis of any group of organisms will usually result in conflicting evidence, i.e. in character incongruence supporting different hypotheses of grouping (Fig. 2). Different characters support different hypotheses of grouping. To take this as a falsification of the theory of descent with modification neglects the distinction between the observation or conceptualization of topological relations of similarity and their interpretation as evidence of common descent. Phylogeny is a unique historical process which cannot be falsified in any strict sense of the word (Popper, 1976b). Order in nature is not predictable in the true sense, since it is not concerned with reproducible processes. Instead it is reconstructed on the principle of connectivity: all topological relations of similarity are potential guides to common ancestry. Character incongruence raises the problem as to which of these characters are to be accepted as indicators of phylogenetic relationship, and which ones simply have to be treated as noise. This was not the case in pre-evolutionary biology, which permitted the acceptance of *multiple affinities*.

It was stated above that the *scala naturae*, a linear arrangement of forms, was based on the principle continuity: if the transition between species or more inclusive types of organization appeared discontinuous to human cognition, continuity had to be postulated on metaphysical grounds (Lepenies, 1978: 44). This at least was the assertion, not only of Leipniz, but also of Bonnet. However, the eager search for transitional types of organization soon produced too many intermediates to be incorporated into an unbranching series of forms. In 1745 already, Vitaliano Donati introduced the metaphor of a net-like arrangement of organisms to account for the multiple affinities of the plants and animals he had studied in the Adriatic sea (Daudin, 1926b: 111; Lepenies, 1978: 45). Even Charles Bonnet had to admit that whereas flying fishes seemed to link the birds with more typical fishes, the latter were also closely approached by a number of marine mammals. Additional equivocations in the linear arrangement of organisms led him to the question: "Does the ladder of life branch in the course of its ascension [to higher levels of organization]?" (Bonnet, 1764, vol. I: 59). The concept of a branching order of nature was further developed by Pierre-Simon Pallas (Mayr, 1982: 209), who even suggested - borrowing from the views of Donati - a polygonal arrangement of organisms to account for their multifaceted affinities (Daudin, 1926b: 164). The great Linné himself, advocate of a rational system of subordinated classification (Callot, 1965) could not deny character incongruence. He equated categorical rank, which he expressed in terms of military hierarchy, with regional zonation (Lepenies, 1978: 48). This led him to the insight that "plants have relationships on all sides as do neighbouring countries on a map of the world" (Mayr, 1982: 175).

The situation did not improve when, under the leadership of Georges Cuvier, the interest of comparative biologists turned away from superficial external resemblances and focused on internal organization. Quite to the contrary, the addition of new characters emerging from the study of comparative anatomy increased the degree of character incongruence. Cuvier, proposing a dichotomously subordinated hierarchy of classification based on logical subdivision (Daudin, 1926a), could not side-step the problem of "*rapports multipliés*" (Daudin, 1926a, II: 109). In his natural history of fishes, published in 1826, he admitted: "It is the inevitable weakness of all methods of classification that they disrupt the multiple affinities for the sake of the most obvious or the most significant ones..." (quoted after Daudin, 1926a, II: 109). The students and collaborators of Cuvier did not fare any better. The crux of all hypotheses of grouping was - and still is - to decide which are the most obvious or most significant affinities, however!

A. Duméril, in his "*Zoologie analytique*" of 1806, undertook to apply Cuvier's method of classification, based on logical subdivision, to the entire animal kingdom. In the course of this project he not only ran into a continuous conflict between the serial and subordinated arrangement or organisms (Daudin, 1926a), documenting the complementarity of these two "ways of seeing" to be further discussed below, but both patterns of classification were furthermore disturbed by character incongruence indicating multiple affinities. This is particularly evident in the case of squamate reptiles which were found to exhibit an almost continuous transition from lizards to snakes, thus rendering the demarcation of the two groups from each other extremely difficult (Oppel, 1811). Viewed from an evolutionary perspective these problems are interpreted as the result of a high degree of convergence within lizards, with many unrelated forms approaching the type of snake habitus and organization to a variable degree. Duméril (1806), by comparison, admitted multiple affinities, not only at the level of squamate reptiles, but also at the level of more inclusive groups - even though he thereby violated his commitment to the Cuvierian "*méthode*": soft-shelled turtles were claimed to exhibit affinities both to crocodiles (then classified together with lizards within the saurians) and to snakes (Duméril, 1806: 82).

Much has been written on W.S. McLeay's *Quinarism*, grouping the taxa in interconnected circles, each circle comprising five groups. The system was obviously highly idealistic, incorporating the Pythagorean enthusiasm for the harmonious and lawful relation of numbers, and it was intended to document the influence of an ordering intelligence responsible for Creation. The search for circles, each composed of five, in the order of nature was an intention popular among many workers of the early 19th century, and whereas it permitted the acceptance of incongruent character distribution, the intention was diametrically opposed to the Darwinian research program. A creationist cannot explain why nature must necessarily be ordered hierarchically (Brady, 1985); Darwin believed to have found the explanation for hierarchical order by his principle of natural selection, as is clear from the only figure included in chapter four of his *Origin*. His theory would never explain circles in five, however. This was enough reason for Darwin and his "bulldog" T.H. Huxley to make the effort to refute this "bizarre" system of classification: viewed with the intention to accept only one unique pattern of subordination of groups within goups, McLeay obviously had to be be charged to have failed distinguishing "affinity" from "analogy" (Mayr, 1982: 202-203). However, the quinarian system was adopted by several pre-evolutionary workers, one of them being John Edward Gray with his "*Synopsis of the Genera of Reptiles and Amphibians, with a Description of some new Species*", published in 1825. The synopsis is structured according to Cuvier's scheme of dichotomization, but this structure could not accommodate character incongruence. Arranging the groups in circles touching each other permitted the expression of multiple affinities of certain key taxa placed at the points of the tangential contact between circles.

In the context of evolutionism, only one single pattern of order in nature would seem admissible. Theory and methods have to do away with character incongruence, i.e. they must permit the choice of one out of all possible hypotheses of grouping. Viewed from an evolutionary perspective, character incongruence must mean that only some characters indicate "true affinity", i.e. historical and genetic continuity of form, while others represent accidental products due to the influence of environmental conditions. In modern terms: only some of the conflicting characters can be interpreted as homologies in a phylogenetic sense, while others must be considered as a result of convergent development, explained by similar selection pressures or similarity of generative mechanisms of form. The problem is to decide which of the many possible hypotheses of grouping are the "most obvious or the most significant" ones, as Cuvier put it. Cuvier's answer to this question was to select that hypothesis of grouping which maximizes the degree of congruence and thus minimalizes the degree of incongruence: the classification supported by the greatest number of congruent characters is the "most obvious one" and hence the one to be accepted. Cuvier did in fact claim to choose the "most significant hypothesis of grouping" based on the subordination of characters according to their physiological importance which he believed he could deduce on an *a priori* basis (Daudin, 1926a, II: 23), but a critical examination of the methodology implied in his work reveals the fact that Cuvier was weighting the characters *a posteriorily* by their congruence, following methodological outlines provided earlier by Antoine de Jussieu (Daudin, 1926b: 208-211). "In sum, the definition of a class now presents itself more overtly for what in fact it has always been: a condensed and coordinated collection of characters revealed by experience" (Daudin, 1926a, vol. II: 66). Outram (1986: 358) noticed the same inconsistency in Cuvier's argumentation: "In the actual process of classification, Cuvier in fact relied ...not on the 'laws' established by the physiological input into natural history, but on the observation and cataloging of minute morphological characteristics...".

This solution, which boils down to an application of the principle of parsimony, is still accepted by supporters of the method of cladistic analysis of relationships (Eldredge and Cracraft, 1980; Wiley, 1981, Nelson and Platnick, 1981), and it is identified by Mayr (1982: 227) as a method of character weighting based on congruence.

If the intention is to represent similarity as a hierarchy of homologies, a reasonable claim to make is that the most complete hierarchy contained within the observed set of topological relations be selected (Brady, 1983). Returning to a theoretical example (Fig. 3), let characters $\underline{a}$, $\underline{b}$, and $\underline{c}$, be incongruently distributed in the taxa $\underline{A}$, $\underline{B}$, and $\underline{C}$. Two alternative hypotheses of grouping (1 and 2) obtain. Looking at these two hypotheses, it is readily appreciated that hypothesis 1) represents the more complete hierarchy contained within the data: characters $\underline{b}$ and $\underline{c}$ appear only once on the branching diagram (the technical term for a 'branching diagram' is cladogram, *clados* being the Greek word for 'branch'), whereas character $\underline{a}$ appears independently in the taxa $\underline{A}$ and $\underline{B}$. Conversely, character $\underline{a}$ appears only once on the cladogram representing hypothesis 2), whereas $\underline{b}$ and $\underline{c}$ make their independent appearance. Information storage is therefore more efficient, or more parsimonious, in hypothesis 1): the information content of the group [$\underline{BC}$] (hypothesis 1) is greater than that of the group [$\underline{AB}$] (hypothesis 2).

Viewed from a phylogenetic perspective, the acceptance of hypothesis 2) implies the explanation of character $\underline{a}$ as an evolutionary novelty, inherited from a common ancestor, while characters $\underline{b}$ and $\underline{c}$ are explained as a result of convergent development. On the basis of this explanation, the hypothesis implies a total of five evolutionary steps (the origin of one novelty plus the evolution of four convergent characters). The provisional acceptance of hypothesis 1) on the other hand implies the explanation of character $\underline{a}$ as an incidence of convergence, while characters $\underline{b}$ and $\underline{c}$ are interpreted as indicators of common ancestry. The hypothesis thus implies a total of 4 evolutionary steps (the origin of two evolutionary novelties plus the development of two convergent occurrences of $\underline{a}$). In sum, hypothesis 1) turns out to be more parsimonious, maximizing congruence and thereby minimizing convergence and hence the number of evolutionary steps involved. Being supported by a larger number of homologies, it is the more "obvious" hypothesis of grouping; as characters b and c show a greater degree of congruence, they are judged to be more reliable (to weigh more heavily or to be more "important") than character a.

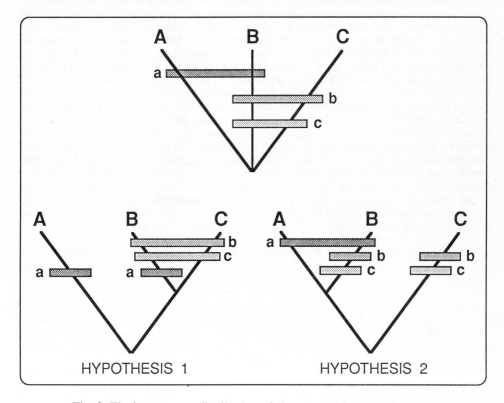

Fig. 3: The incongruent distribution of characters a, b, and c in the taxa A, B, and C supports two alternative hypotheses of grouping, 1) and 2). Of the two hypotheses, 1) is more parsimonious, as it maximizes congruence and thereby minimizes convergence.

It is at this juncture that the distinction of topographical from phylogenetic relations of similarity, as discussed in the preceding paragraph, becomes important. Character incongruence can only be a problem with respect to features which are in accordance with the topographical criteria of similarity. Would the characters in question differ in their topological relations, there would be no problem to distinguish homology from convergence by induction, i.e. on observational grounds. The very fact of character incongruence shows, however, that hypotheses of common ancestry must derive from the evaluation of alternative hypotheses of grouping on the basis of parsimony (Rieppel, 1980). The methodological criteria of comparative biology differentiate between topographical correspondence and non-homology. Similarity of form and function in non-homologous relations is termed analogy in contemporary biology. Topographical relations of similarity correspond to the "*analogies*" of E. Geoffroy Saint-Hilaire, and to the homologies of non-evolutionary comparative anatomy as expounded by Richard Owen.

In the context of evolutionism, topological correspondence functions as a potential guide to common ancestry. In the case of character congruence, the topographical relations of similarity are interpreted as homologies, which implies the hypothesis that the features in question can be traced back to corresponding features in the common ancestor of the taxa under analysis. By this interpretation, the term "homology" changes its meaning as compared to pre-evolutionary usage. Before the Darwinian revolution it denoted static relations of coexistence. When Darwin explained the "unity of type" by "unity of descent", the term "homology" came to denote dynamic relations of succession as it assimilated the idea of common ancestry.

At an early date already, Lankaster (1870) recognized the dilemma created by the dual meaning of the notion of homology. To avoid conceptual confusion he suggested to distinguish homologous from homogenous relations of similarity, i.e. homology from homogeny. Homology in Lankaster's (1870) use denoted topographical correspondence and thus preserved the pre-evolutionary meaning it had in the writings of Richard Owen: "the same organ in different animals under every variety of form and function" (Owen, quoted from Lankaster, 1870: 35). If homologous structures are considered to be "genetically related, in so far as they have a single representative in a common ancestor, [they] may be called homogenous" (Lankaster, 1870: 36). However, "when identical or nearly similar forces, or environments, act on two or more parts of an organism which are exactly or nearly alike, the resulting modifications of the various parts will be exactly or nearly alike"; Lankaster (1870: 39) proposed to call this kind of agreement *homoplasis* or *homoplasy*. A close analysis of his text shows that his term homoplasy covers both convergence as well as parallel evolution, which was expressed by Wiley (1981: 12; emphasis added) as follows: "Homoplasies are characters that display structural (and thus ontogenetic) similarities but are *thought* to have originated independently of each other, either from two different pre-existing characters or from a single pre-existing character at two different times or in two different species". The reason to *think* so is character incongruence.

In an earlier attempt to discuss the same problem, Rieppel (1980) distinguished "topographical homology" (topographical correspondence) from "phylogenetic homology". This choice of terms was inadequate, since by it convergence became a category of homology. The discussion of Lankaster's (1870) paper shows "topographical homology" to correspond to the pre-evolutionary meaning of the term "homology", whereas "phylogenetic homology" turns into "homogeny". It would indeed appear advantageous to return to Lankaster's (1870) clear-cut terminology, were it not for the universal use, in modern biology, of the term "homology" in an evolutionary context. The terminology used in the present text is tabulated below.

The Relation of Homology and Homoplasy			
	topological relations of similarity	character congruence	character incongruence
Lankaster, 1870	homology	homogeny	homoplasy
Modern Biology	topographical correspondence	homology	homoplasy (convergence parallelism)

Homology as used in the present context is synonymous with synapomorphy, i.e. with the occurrence of shared derived characters corroborating the hypothesis of monophyly in a cladistic analysis (Patterson, 1982). It has been argued (Ax, 1984: 183) that the synonymy of homology with synapomorphy is based on faulty logic, since symplesiomorphies may also be homologous features. However, this criticism does not appreciate the dual meaning of homology, as a topographical relation of similarity and as an indication of common ancestry, and it is based on a misapprehension of the hierarchical nesting of homologies which is congruent with the genealogical hierarchy as will be more fully discussed below (Riedl, 1975; Eldredge, 1985). Both, synapomorphies and symplesiomorphies correspond to topological relations of similarity, but only synapomorphies provide a guide to common ancestry. Secondly, every synapomorphy turns into a symplesiomorphy at a less inclusive, i.e subordinated level of the hierarchy: the tetrapod limb is a synapomorphy at the level of the Tetrapoda, but it is a symplesiomorphy at the level of the Amniota, a less inclusive group nested within the Tetrapoda .

In conclusion then, phylogenetic reasoning follows a deductive scheme as illustrated in Fig. 4. The "*principe des connexions*" discriminates between topological relations of similarity and non-homology: the latter characters fail the test of similarity (Patterson, 1982). Only topographical correspondence can be considered as potential guide to ancestry. If topological relations of similarity are incongruently distributed, they support alternative hypotheses of grouping. In other words: not every topographical correspondence can be interpreted as a homologous relation of similarity, to be explained on the hypothesis of common ancestry. The test of congruence (Patterson, 1982) is applied to discriminate between homology and homoplasy.

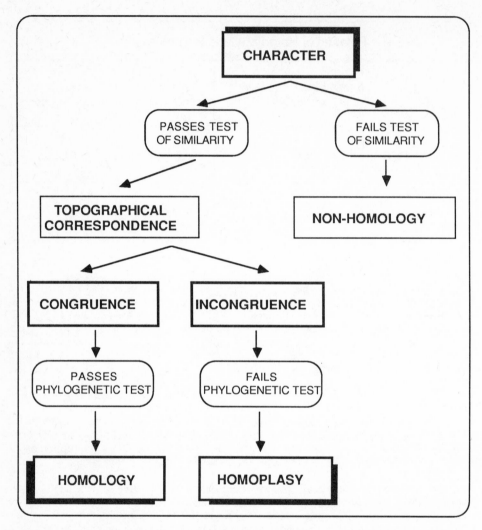

Fig. 4. The relation of homologous similarity (synapomorphy) to non-homology or homoplasy (convergence).

Congruence versus incongruence are evaluated on the basis of parsimony: this introduces another methodological rule, resulting from the intention to seek for the most complete hierarchy contained within the observed set of topographical relations of similarity (Brady, 1983). As was discussed in the first chapter, the principle of parsimony originated with William Ockahm (ca. 1280-1349), who distinguished between the notion based on the *act* of perception, and the *judgment* as to whether the object perceived or designated by the notion has indeed material reality or not (Imbach, 1981). A fundamental uncertainty therefore dominates human cognition: Ockham solved the problem by the introduction of the parsimony principle.

Burrowing from Ockham it might be stated, in the context of modern comparative biology, that the act of perception, or conceptualization, of topological relations of similarity (based on the principle of connectivity) must be distinguished from the judgment as to whether these relations of similarity indicate common ancestry (on the basis of congruence, i.e. parsimony). Those characters which show the greatest degree of congruence are interpreted as evolutionary novelties, inherited from a common ancestor, whereas incongruent characters are explained by convergent evolution. It is impossible to *know* whether this explanation, or interpretation, of observed similarity is the correct one - but it must be accepted as the most *likely* one (Sober, 1983). All knowledge is hypothetical and never transcends the epistemological limitations of the perceiving object; every hypothesis remains nothing but a "way of seeing", a view of the world framed within the categories of human thought or experience, and dependent on theoretical premises and methodological rules.

It must be borne in mind that the parsimony principle is an intended methodological rule, an imperative which bids and forbids: it is not an argument for objectivity or empiricism! The application of this principle in phylogeny reconstruction seems to be unavoidable: resulting from the intention to find the most complete hierarchy in the given data set (Brady, 1983), it provides the appropriate tool to deal with character incongruence. One might attempt to select the most "significant" hypothesis of grouping on the basis of criteria of character weighting instead of congruence, i.e. by criteria which assign some kind of "physiological importance" or "phylogenetic inertia" to the characters in question. It has been mentioned already that Cuvier purported to classify organisms on the basis of the subordination of characters according to their functional importance which he claimed to be able to assess by "physiological deductions" (Daudin, 1926a, II: 23). In a similar vein, Hecht and Edwards (1977: 15-16) have designed criteria of character weighting which attribute more informational content to "those states which are part of a highly integrated functional complex" than to a "simplification or reduction" of characters. The rationale is that a functionally integrated character complex has a more complex genetic basis, and is therefore less likely to undergo convergent change, than the loss, reduction or simplification of structures. The problem, however, is that there is no unequivocal measure of complexity available at the present time, nor can the assumptions implicit in these criteria of weighting be justified by the present state of development of evolutionary theory (Gaffney, 1979; Brady, 1982). The criteria are furthermore based on a faulty epistemology, since they presuppose assumptions about the course of the phylogenetic process which, in fact, can only itself provide hypothetical explanations of the results of phylogeny reconstruction. To avoid circular reasoning, the latter should proceed independently from any prejudice concerning the likelihood of changes to occur (Rosen, Forey, Gardiner and Patterson, 1981; Brady, 1982; 1985). As Patterson (1982) has put it: to recognize a character is to recognize a group. Weighting of characters therefore cannot proceed independently of a hypothesis of grouping and hence of relationship, or, to put it in other words: characters weight themselves, with those forming useful groups (in congruence with other characters) carrying more information content than the others.

The admission of the principle of parsimony into the methodology of phylogeny reconstruction is thus an acknowledgment of ignorance. This ignorance concerns the supposed "truth" of the phylogenetic process: not only is this unavoidable from an epistemological point of view, but it also results from the simple fact that phylogeny has occurred in the distant past and therefore is no longer amenable to observation. Viewed from a pragmatic perspective, comparative biology profits from the application of the parsimony principle: since it maximizes character congruence (and by this minimizes incongruence), the information content of formal classifications is also maximized (Nelson and Platnick, 1981).

A Note on Molecular Systematics

Recent years have witnessed a spectacular rise in molecular systematics which is given preference over the comparative morphological approach to phylogeny reconstruction for a number of reasons. Protein structure, and particularly amino acid sequences, are believed to reflect the genetic makeup of the organisms more closely than their anatomy, and hence to carry a greater phylogenetic information content. This is expected to result in a lesser degree of character incongruence, giving rise to more robust hypotheses of phylogenetic relationships.

The bearing of molecular data on phylogeny reconstruction was the topic of a recently published symposium (Patterson, 1987). Opinions on the prevalence of molecular data over comparative morphology were divided, ranging from strong support to cautionary mitigation of uncritical enthusiasm.

For one thing, the problems of homologization which haunt comparative anatomy do not disappear with a molecular approach (Raff et al., 1987: 219). McKenna (1987: 58) cautioned: "...molecular studies can suffer from exactly the same ills that beset comparative anatomical ones: a touchstone has not been found that will test phylogeny unequivocally". The existence of multiple copies of molecular homologs (for which the technical term is *homonyms*) results in a statistical view of molecular homology (Patterson, 1977). Problems of the explanation, in phylogenetic terms, of similarity at the molecular level are aggreviated by the fact that molecular data are strictly phenetic (Andrews, 1987: 29), so that relations of similarity have to be established on a statistical basis (Sibley & Ahlquist, 1987: 100; Goodman et al., 1987: 147). For that reason, molecular systematics come close to relying on "overall similarity" as a guide to common ancestry, a striking departure from cladistic methodology which focuses on "special homology" (synapomorphy) for phylogeny reconstruction. The analysis of molecular data approaches phenetic taxonomy (Andrews, 1987: 209), or at best it represents "a dubious mixture of phenetic and cladistic methodology" (McKenna, 1987: 37). A further obstacle in the comparison of molecular data results from exon duplication and shuffling, a process which may create hybrid genes that can only be compared in terms of "partial homology": this is a striking contrast to the morphological approach which treats homology as an all-or-none relation (Patterson, 1977; see also Gould, 1988).

The most interesting, and most controversial, tenet of molecular taxonomy is that its methodology should be less prone to character incongruence. Goodman et al. (1987: 147), in particular, favor molecular data as opposed to morphological data, even if "simple statistical procedures", revealing the number of matching nucleotide bases, are required to establish relations of homology. The data thus obtained can then be subjected to some powerful maximum parsimony algorithm which will yield the tree with the smallest number of base changes. If molecular data, treated in the appropriate manner, were indeed superior to morphological comparison, one would expect a higher degree of character congruence yielding more fully resolved and more robust cladograms.

This claim has been tested in a recent paper by Wyss, Novacek and McKenna (1987), with a negative result. The authors compared the resolving powers of newly acquired morphological data as opposed to recently published amino acid sequence data in an analysis of mammalian interrelationships at the ordinal level. In view of the current trend to concentrate on molecular studies, the results of these authors are highly interesting, because morphology proved to be of superior resolving power: "We are not discounting the importance of sequence-based systematics but in fact welcome it as a useful test of morphology...at present, morphology remains the most powerful comparative tool for identifying eutherian orders and their mutual affinities" (Wyss *et al.*, 1987: 111). In other words, molecular data did not fare any better with respect to character incongruence than morphology: "amino acid-sequence techniques are no more secure than are those based on comparative anatomy (Wyss *et al.*, 1987: 111).

The rise of molecular systematics, the use of amino acid sequencing in phylogeny reconstruction, did not offer any solution to the basic problem faced by every systematist, namely to seek "a resolved pattern for incompatible sets of apparent homologies" (Wyss *et al.*, 1987: 110). The "apparent homologies" are topological relations of similarity in amino acid-sequences which behave as incongruently as morphological characters, or even more so. The test of homology at the molecular level must again be provided by congruence, and therefore relates to the principle of parsimony.

A Note on Parallel Evolution

A distinction is frequently made between parallel evolution and convergence. The latter denotes the independent evolution of similar characteristics in two or more unrelated lineages. Parallelism on the other hand implies the independent evolution of similar features in two lineages descending from a common ancestor which itself is assumed to lack this feature. The argument that evolutionary novelties should arise independently, i.e. in parallel in two related lineages is based on the assumption of a similar genetic background, inherited by these descendent lineages from their common ancestor. It is further argued that because of their similar genetic makeup, the lineages would respond in an equivalent manner to similar selection pressures.

The concept of parallel evolution is based on mistaken epistemology, because it presupposes knowledge of phylogeny in the reconstruction of pattern (Schoch, 1986). Similarity is explained neither as homology (i.e. as a character inherited from a common ancestor), nor as an incidence of convergence (independently evolved similarity), but as the result of independent evolution in two related lineages: knowledge of the relationship of these lineages is implicit in the call for parallelism, but cannot obtain from the data. What is observed are topographical relations of similarity. These can be interpreted either as evidence of phylogenetic relationship, or as convergence. Parallelism can only be called for on the basis of a given hypothesis of grouping (or phylogeny) as it obtains from the analysis of other characters, and by the addition of the *ad hoc* hypothesis that the common ancestor did not share the similarity observed in its descendants (Eldredge and Cracraft, 1980). Patterson (1982: 50) justly concluded "that parallelism...is nothing but a hindrance to phylogenetic analysis".

The Test of Homology

"Classification is an information storage and retrieval system" (Mayr, 1969: 229). The application of the parsimony principle maximizes the information contained within the hierarchy of homologies and therewith in the genealogical hierarchy as long as the latter is congruent with the first. Information storage and retrieval is also maximized for classifications mirroring the hierarchy of homologies. A more interesting aspect of classifications is their *predictive value* with respect to the information stored: if homologies turn out to be predictable, they would be open to test in a Popperian sense. Homologies are indeed claimed to be predictable on the basis of current knowledge of order in nature:

should a new mammalian species be discovered, it can be predicted to have hair, mammary glands, a single bone in the lower jaw, a left aortic arch, etc. A whole suite of characters thus appears to be predictable. Should a mammal be discovered which lacks some or all of these characters, the discovery would falsify the current concept of the Mammalia.

It is a consequence of character incongruence that organisms are discovered and described which violate the diagnoses of known groups to a greater or lesser degree. In practice, two options obtain to deal with such a situation (Rieppel, 1983a). Either the group in question is considered to have been ill-defined, and the criteria of membership are revised so as to include the new candidate, or else, the diagnosis of the existing groups is left unaltered, and a new group is erected to include the candidate in question. The point is, however, that these changes of grouping or of diagnosis, thought to falsify predictions of homologies, really correspond to rationalizations after the fact! It is pattern analysis which reveals a greater or lesser degree of congruence in the first place; the prediction of homologies must be based not on the recovered pattern, but on a causal theory which explains *why* homologies should at all behave congruently and hence be predictable, and it is this theory which is falsified by character incongruence. Such a theory may relate to the developmental and/or physiological (functional) correlation of characters (see chapter 6).

That the "predictions" of homologies based on classification are rationalizations after the fact (Coleman, 1961: 121) can be illustrated by historical as well as by modern examples. Cuvier, for instance, emphatically predicted the occurrence of marsupial bones in an incompletely prepared fossil from the quarries of Montmartre which he recognized as a member of this group on the basis of other characters. His prediction was indeed corroborated by the complete preparation of the pelvic girdle, but as Coleman (1961: 203, footnote 17) noted, it really boiled down to an "inspired guess" since these bones may be reduced in some marsupials. Outram (1986: 359, footnote 107; emphasis added), in her discussion of the same incidence, stressed that "such a demonstration...showed the power of systems of *observed correlations,* rather than true predictive power". On the other hand, marsupial bones are also observed in monotremes and in fossil multituberculates.

T.H. Huxley, in a review of the fossil mammalian genus *Macrauchenia*, voiced a word of caution in his comments on R. Owen's earlier work on that genus. Highlighting character incongruence, Huxley discussed the possibility to predict homologies on the basis of incomplete specimens, only to conclude: "But, for all that, our hypothetical anatomist would have been wrong; and instead of finding what he sought, he would have learnt a lesson of caution, of great service to his future progress" (quoted from Rachootin, 1987: 177).

A modern example derives from the study of limbless scincid lizards from South Africa, belonging to the subfamily Acontinae, family Scincidae (Rieppel, 1982). The subfamily comprises three genera, *Acontias, Acontophiops,* and*Typhlosaurus,* which form a monophyletic group (Greer, 1970). The acontine skinks are sand-dwelling, fossorial lizards whose mode of life requires a protection of the eye from the infiltration of grains of sand: three stages of progressive modification are observed. The genus *Acontias* retains movable eyelids, and thereby represents the generalized condition. *Acontophiops* retains an eyelid, but it is fused to the surrounding skin of the head along the anterior, ventral and posterior border of the eye; only its upper edge remains free, i.e. unfused. *Acontophiops* therefore represents an intermediate stage of modification. In *Typhlosaurus*, the eye is completely covered by a scale of the skin covering the head. The eyelid has been lost or modified, the most advanced condition.

Modification of the eye and its adnexae is not the only adaptation of these lizards to a sand-dwelling mode of life. Selection pressure resulting from burrowing habits favors a diminuation of the diameter of the head, what results in the miniaturization of the skull in these (and similar) lizards (Rieppel, 1984b). One of the many morphological modifications resulting from miniaturization of the skull is the loss of the upper temporal arch in most members of the Acontinae (Rieppel, 1982). The occurrence of an upper temporal arcade is a generalized feature within lizards; one possible mechanism explaining its loss is paedomorphosis (Rieppel, 1984c), i.e. the arrested development of the bones constituting the upper temporal arcade. Interestingly enough, the genera *Acontias* and *Typhlosaurus* each include one single representative which primitively retains the upper temporal arcade (Greer, 1970). In view of its intermediate position with respect to the structure of its eye, the prediction would be that *Acontophiops*, a monotypic genus, retains the upper temporal arcade. Such is, however, not the case (Rieppel, 1982). The characters in question obviously change independently from each other. This phenomenon, which is also observed with respect to other features in different organisms, has been called *mosaic evolution* by deBeer (1954; see also Rieppel, 1979a), and it explains the occurrence of character incongruence.

Patterson (1982: 58) discussed various sources from which conjectures of homology might emanate: perhaps the most interesting argument is that homologies can be treated as conjectures "whose source is immaterial to their status" (Patterson, 1982: 58). Their status is evaluated with respect to their success or failure to pass a series of tests. Three methods of testing have been proposed by Patterson (1982: 37-38) which will be discussed in sequence: similarity, convergence, and conjunction.

Patterson's test of similarity includes criteria of topological relations and of ontogeny. The test of topological relations boils down to the application of the principle of connectivity as a guide in the observation of similarity which abstracts from specific form and function of organs and organ systems. As was detailed above only those characters which pass the test of similarity are of potential interest in phylogeny reconstruction.

The test of conjectured homology by ontogenetic sequence will be dealt with in greater detail in the chapters to follow. Suffice it to say at this juncture that the *phylogenetic* interpretation of ontogenetic stages (the term as such entails a problem already, because so-called "stages" have to be abstracted from a continuous developmental process: Alberch, 1985; Rieppel, 1985c) is not independent from the hierarchy of homologies as determined by congruence (Rieppel, 1979b; Fink, 1982; Kluge, 1985; Nelson, 1985). This is a point to be discussed in greater detail under the heading of phylogeny reconstruction by ontogeny (chapter 6). The test of ontogeny relies on von Baerian recapitulation, stating that the more general character (or character-state) precedes the less generally distributed character (or character-state) in development. Congruence of independent characters may indicate in a given instance, however, that the occurrence of a more generally distributed character (or character-state) is more parsimoniously explained in phylogenetic terms as an incidence of character reversal (by paedomorphosis) than as a truly generalized condition.

This brings us to the discussion of Patterson's (1982) second test of homology, i.e. congruence. Although Patterson (1978) had previously argued that the test of homology by congruence invokes no more than the principle of parsimony, he changed his view in the paper of 1982. In contrast it was argued above that the application of the parsimony principle amounts to a methodological rule. On the whole, the choice of the "most simple" hypothesis of grouping boils down to a conventional agreement (Popper, 1976a: 105), though a necessary one. Linking the parsimony principle to some measure of probability does not help either to render it a test in the strict sense of the word. Although the genealogical explanation of the most parsimonious hypothesis of grouping is the most likely one, there is no logical necessity to assume that genealogy must exclude homoplasy (Sober, 1983).

It must furthermore be stressed that the test by congruence can only be applied to characters which are in accordance with the methodological (topographical) criteria of homology, if the procedure is not to be caught in a circle. In other words: congruence does not test topographical correspondence; rather it tests the interpretation of topological relations of similarity as a guide to common ancestry (Fig. 4). An example may serve to illustrate this point. In 1982, Gardiner published his highly controversial paper on the classification of tetrapods in which he proposed, *inter alia*, a sister-group relationship between birds and mammals. This proposal goes back to pre-evolutionary biology, and was also defended by Richard Owen in 1866 (Gardiner, 1982: 209-211). Ever since, the classification of birds and mammals within a monophyletic group termed Haemothermia (the name alludes to the endotherm physiology typical of both groups) seemed to contradict all that was learnt during subsequent studies of comparative anatomy (Goodrich, 1916) and paleontology (Kemp, 1982, 1987). In support of this controversial hypothesis, Gardiner postulates the mammalian alisphenoid bone to be homologous with the avian orbitosphenoid bone, the two elements really representing the reptilian epipterygoid.

The nature and position of the reptilian epipterygoid was discussed above, and it was noted that this bone has changed its topographical relations to the branches of the trigeminal nerve in the amphisbaenian genus *Trogonophis*. In view of the conceptual problems involved with topographical correspondence, Gardiner (1982) decided to rely more heavily on congruence for the establishment of homology. He concluded that reptiles, birds and mammals all share a homologous bone in the orbitotemporal region of the skull, the original epipterygoid, which becomes included in the lateral braincase wall in mammals and birds, there forming the alisphenoid (mammals) or pleurosphenoid (birds), respectively. The cavum epiptericum, lying deep to the epipterygoid, would thus have become secondarily incorporated into the cranial cavity. (A convergent development of a similar condition is postulated to have occurred in snakes, but see Rieppel, 1976, for a different interpretation of the orbitotemporal region of the snake skull.) In Gardiner's view the changing topographical relations of the bone to the branches of the trigeminal nerve cannot refute his conjecture of homology: the element is in a seemingly comparable position with respect to the nerves in *Trogonophis*, mammals and birds (and snakes)!

With his theory, Gardiner (1982) contradicts all textbook interpretations of the avian skull. While this in itself cannot constitute an argument, the topographical relationships of the bones in question must be reconsidered. It is widely accepted that the mammalian alisphenoid incorporates the lacertilian epipterygoid (Presley and Steel, 1976). The latter, however, is of splanchnic origin, representing a derivate of the palatoquadrate. It is therefore not an original element of the lateral braincase wall, but has secondarily become incorporated into the braincase wall in mammals. The same is not true for the orbitosphenoid of birds, which is considered to be an ossification within the primary lateral wall of the braincase, originating from the pila antotica (deBeer, 1937). The reptilian epipterygoid is lost in birds, as well as in crocodiles, which also show a pleurosphenoid ossification. In conclusion, the topograpical relations of the pleurosphenoid and alisphenoid to the primary lateral braincase wall is different, which is why the homology of the two ossifications is disputed in textbooks. Gardiner (1982) might argue that the determination of relations of topography has simply exchanged the branches of the trigeminal nerve for the primary lateral wall of the braincase as fixed frame of reference, but that no evidence has in principle been produced to show that this new frame of reference is more reliable than the course of nerve branches. On the other hand there is, according to

Gardiner (1982), a large suite of congruent characters supporting the sister-group relationship between birds and mammals, and if this hypothesis of grouping is accepted on grounds of parsimony, it permits the homologization of the mammalian alisphenoid with the avian (indeed archosaurian) pleurosphenoid on the basis of congruence. A denial of his conjecture of homology is judged to be unparsimonious by Gardiner (1982: 215).

Turning to the other characters purported to group mammals with birds, one finds a number of features which are conceivably related to endothermy. One might intuitively judge these characters to represent the correlated result of the convergent development of endothermy in the two groups in question. Admittedly, intuition should not play a role in the evaluation of scientific hypotheses (although it frequently does). But if homologies can be considered to represent testable conjectures the material source of which is irrelevant, it must be requested that the conjecture of the homology of endothermy in mammals and birds be tested. Gardiner (1982) would have to test the hypothesized homology of warm-bloodedness in the Haemothermia by the demonstrated congruence of other, independent features. One of these would have to be the alledged homology of the mammalian alispehenoid with the avian pleurosphenoid. The argument would thus have closed the circle.

Congruence or parsimony alone, with no consideration of topographic relations, would permit the homologization of any character with any other: there is no criterion to demarcate homology from non-homology outside a circular argument. In other words: incongruence falsifies the conjecture of homologies as guide to common ancestry; it refutes the phylogenetic interpretation of topographical correspondences, but not topological relations of similarity in themselves. Pattern reconstruction must proceed by the recording of topological relations of similarity, which all are potential guides to common ancestry. Congruence and parsimony are the test and the tool to select those similarities which are the best (most likely) candidates as indicators of common descent. Kluge and Strauss (1985: 258) have called congruence the "final test" of homology, as opposed to similarity (topographic, compositional, and ontogenetic) which is considered as a preliminary test only. This view is in accordance with their understanding of homology as a guide to common ancestry (Kluge and Strauss, 1985: 257): their "final test" is one of potential phylogenetic significance, not of topological relations of similarity.

How could it be possible to refute Gardiner's (1982: 215) contention that the epipterygoid of reptiles, the pleurosphenoid of crocodiles and birds, and the alisphenoid of mammals are all homologous? This claim can in fact be refuted, at least as far as crocodiles are concerned, namely by Patterson's third test of homology, the test of conjunction: "the homologues may not coexist in one organism" (Patterson, 1982: 21). The epipterygoid was introduced above as an ossification of the ascending process of the palatoquadrate: although this statement implies the prior identification of the palatoquadrate on independent criteria, the process of ossification is amenable to observation in a series of developmental stages. The pleurosphenoid of crocodiles and birds, on the other hand, was introduced as an ossification of the primary lateral wall of the braincase, i.e. of the pila antotica. The homology of the two ossifications can be refuted if the existence of both structures is demonstrated in one organism. This is the case in crocodiles (Bellairs and Kamal, 1981: 238) where a cartilaginous primary lateral wall of the braincase (pila metoptica and pila antotica), eventually ossifying as pleurosphenoid, occurs in conjuction with a rudimentary ascending process of the palatoquadrate which makes its appearance during some stages of development only.

Conjunction appears to be a powerful test of homology quite independent of additional criteria such as conventionally accepted operational rules, but it may fail if the whole life cycle of animals with an ontogeny is taken into account (Patterson, 1982). Thus, the presence versus absence of a character cannot be tested "since most homologies present later in life are absent in the zygote" (Patterson, 1982: 48). Furthermore, the test of conjunction cannot be applied in cases of ontogenetic transformations (Haeckelian recapitulation). Patterson (1982: 52) notes that "ontogeny does some time show transformations from one condition to another", and as an example quotes the migration of one eye in flatfishes (Nelson, 1978), or the secondary closure of originally open visceral clefts. In this case, the two homologous conditions, open versus closed visceral clefts, are consecutively present in one and the same organism, and the character would fail the test of conjunction if the latter were applied uncritically. Another example demonstrating the same problem is more fully discussed under the heading of objectification of the genealogical hierarchy by homology (chapter 5). It concerns the replacement of the primary by the secondary lower jaw joint during a postnatal stage of development in some marsupials (Starck, 1979: 341). That the test of conjunction cannot account for ontogenetic transformations must again be recognized with respect to other, operational, criteria of homology.

Tested Intuition?

Whether or not Belon's comparison of the skeleton of a bird and man was guided by intuition, whether or not the classification of natural groups within groups is structured by cognitive mechanisms: what is essential for comparative biology is the epistemological limitation that characters cannot be "discovered" in any material sense. Patterson (1982: 58) has pointed out that the material basis of a conjecture of homology may be considered immaterial to its status if ways are found to subject these conjectures to test and potential refutation. Three such tests were proposed by him, viz. the tests of similarity, of congruence and of conjunction. Here, the point has been made (Fig. 4) that the test of congruence is subordinated to the test of similarity.

To be more explicit: similarity can be claimed to be observed, recorded, or conceptualized on the basis of the *principe des connexions*, which is a guide to abstraction from the particular form and function of organs or organ systems. It might also be claimed that similarity is a bold conjecture, the material basis of which is irrelevant to its status, but which must be testable and potentially falsifiable in order to be empirical. The test then is that of topological relations, rather than of specific form and function. The recognition of a character constitutes a conjecture of similarity, establishing a relation between appearances or, in a more practical sense, it corresponds to a hypothesis of grouping. The test of similarity differentiates between biologically meaningful classifications and such meaningless groups as "red objects". The character in question must first be subjected to the test of similarity in order to be considered as a potential guide to common ancestry.

It might be argued that this argument is irrelevant, because meaningless hypotheses of grouping, based on biologically irrelevant characters, would be eliminated by the test of congruence. I maintain, however, that the test of congruence, if applied to a data set with no support from additional criteria, is circular. The test of congruence can only be applied to characters which satisfy the topological criteria of similarity. Congruence does not test topographical correspondence versus non-homology, it tests homology versus homoplasy. Conjunction is a powerful test of homology which can be applied at all stages of analysis except in cases of developmental transformations. The limitations of the tests of congruence and of conjunction can only be appreciated with other, independent criteria of homology, which in turn derive from E. Geoffroy Saint-Hilaire's "*principe des connexions*".

CHAPTER 4

ONTOGENY AND HIERARCHY

The comparison of animals is based on the concept of homology. The conjecture of relations of homology presupposes the decomposition of organic structures into constituent elements, and the comparison of the latter in terms of topographic relationships. The constituent elements of organic structure must be created by a continuous developmental process. Beyond the point that a series of discontinuous stages must be abstracted from the continuous process of ontogeny in order to obtain stages or characters suitable for comparison (Alberch, 1985; Rieppel, 1985c), some kind of relationship is to be expected between the hierarchy of homologies and the hierarchy of the ontogenetic process of differentiation. In other words: to the extent that the process of ontogenetic differentiation is ordered, this order will be reflected in the hierarchy of homologies created by development. The structure of the hierarchy of homologies, however, may differ with different models of ontogeny being accepted.

<u>Recapitulation</u>

The idea of a parallelism between the gradation of organic complexity, as implicit in the concept of the *scala naturae*, and the process of ontogenetic development can again be traced back to Aristotle, who not only linked the idea of a gradual increase of perfection or organization to the principle of continuity, but who himself investigated the development of a chick. Was it not surprising that during early developmental stages the chick would resemble a maggot, as William Harvey was to confirm in 1651? Aristotle concluded that all creatures, even the most perfected ones, are produced from initial worm-like stages. The same analogy persisted into Albrecht von Haller's (1758) study of the development of the chick embryo, as well as into Charles Bonnet's *Palingénésie* (1769, vol. I: 178): "...these strange revolutions which the chick undergoes from the moment at which it first becomes visible until the time at which its true form becomes apparent. I shall not recapitulate these revolutions at this point: I shall confine myself to remind my readers of the fact that as the chick embryo starts to become visible, its form appears to closely approach that of a very small worm". In his *De Motu Cordis* Harvey (1628) went even further, specifying that in the early chick embryo the heart makes its first appearance in the form of a simple vesicle and thus resembles the adult structure of the heart in animals of lower perfection. This probably represents the first specific statement of recapitulation of a certain organ, during its development, of stages of differentiation of lower perfection.

For Aristotle, the hierarchy of somatic organization corresponded to a hierarchy of souls, while the perfection of the body mirrored the additive functions of these souls. Aristotle distinguished three types of souls: the vegetative or plant soul was active in all living beings. The sensitive or animal soul, however, was present in animals and in man only, while the rational soul was restricted to man. Aristotle held that during ontogeny the human being would successively actualize the functions of this hierarchy of souls in correlation with the successive development of the body, so that the two substances may work together in harmony. During early developmental stages, the vegetative soul alone would be active, providing for the vegetative functions, i.e. for the nutrition of the embryo. The latter was viewed to be rooted in the mother's womb by means of the umbilical cord. Later, the embryo shows signs of sensitivity and movement. The stage is reached at which, according to Albrecht von Haller (1755), the muscles have acquired enough consistence to

69

become "*irritable*". The animal soul has become functional. The rational soul is not put into action until postembryonic development, which explains why the behavior of the baby appears so strikingly animal-like.

This Aristotelian theory of development is the matrix upon which Charles Bonnet (1769) built his *Palingénésie Philosophique*. Bonnet was among the first to present the Great Chain of Being in its temporalized version which he derived from Leibnizian metaphysics (Lovejoy, 1936). As mentioned above, he postulated that divine foresight had programed a series of earthly revolutions which would trigger the development of preformed 'germs of resurrection' into evermore perfected types of organized beings. That way the immortal soul would ascend all rungs in the ladder of life, and since in his view the soul was intimately and reciprocally related to the body, the increasing (but preformed!) perfection of the body would permit an increasing perfection of the receptions of the soul. A consequence of this view of life was that the developing germ was interpreted to recapitulate his whole "history", i.e. his ascent through the *scala naturae*, and the inspection of the "*métamorphoses*" or "*révolutions*" during his "*évolution*" might thus not only provide a key to the past, but might even permit to catch a glimpse of the future (Bonnet, 1769, vol. I: 184). "Does infancy not correspond to a stage of pure animality...?, asked Bonnet (1769, vol. I: 288), and drew form this observation the logical conclusion: "The animals are today in the stage of infancy; perhaps they will one day reach the stage of a thinking being...". Darwin, in his B-notebook of 1837 (Gruber and Barrett, 1974: 158, 213), entered: "If all men were dead, then monkeys make men. - Men make angels". His notion of "evolution" meant something entirely different, however. Where Bonnet was talking about the unfolding of a preformed germ, Darwin implied the development of something entirely new by descent with modification.

The theory of recapitulation of lesser stages of perfection during the ontogeny of a more perfected being can thus be demonstrated to be derived from the Aristotelian concept of a gradation of organic complexity paralleled by the gradation of mental abilities. Ontogeny was interpreted as to recapitulate the *scala naturae*, demonstrating the ascent of life through all the rungs of the ladder of complexity. If, in expansion of this retrospective view, ontogeny is made the mechanism of progressive development, it must logically follow that an increase in perfection of organization is achieved by the *terminal addition* of new developmental stages to the ancestral ontogeny. This may be viewed as a consequence of divine preformation, as in the case of Charles Bonnet, or progressive development may be reduced to natural mechanisms as it was done by E. Serres and J.F. Meckel. A point to note is that 18th century theories of recapitulation postulated the whole organization lower organism to be recapitulated during the ontogeny of higher organisms: ontogeny litterally ascended the ladder of life. This is an important difference from the more sophisticated theories, which centered on the development of single organs or organ complexes, i.e. of constituent elements rather than of the whole organism, in support of recapitulation.

The Meckel-Serres Law of Terminal Addition

"There is no competent physiologist who would not have been struck by the remark, that the original form of all organisms is one and the same, and that from this one form all others, the lowest as well as the highest, develop in such a way that the latter pass through the permanent forms of the first as transient stages only. Aristotle, Haller, Harvey, Kielmeyer, Autenrieth and others have remarked this either in passing, or they have, as the latter authors in particular, insisted on this fact and deduced from it unforgettable results for physiology": these are among the opening remarks of J.F. Meckel to his *Contributions to Comparative Anatomy* (1811: 3). Carl Friedrich Kielmeyer is of particular importance among the authors quoted by Meckel, since he enunciated the Aristotelian concept of a hierarchy of souls paralleled by the developmental stages of the body in his address given on the occasion of the birthday of the duke Carl von Wirtemberg on 2 February 1793, therewith setting the stage for the development of *Naturphilosophie* in Germany (Russell, 1916), which was to influence the French transcendentalists (see below).

Noteworthy in the quote from Meckel given above is the adherence to the concept of the *scala naturae*. His treatise on comparative anatomy reviewed the development of a number of organs or organ-complexes in order to demonstrate the fact of recapitulation, which in turn was used to prove the parallelism existing between the human fetus and the "animals standing below it" (p. 60). An assertion such as this presupposes the concept of homology, since a comparison is implied between embryonic stages of man and adult stages of less perfect animals. As discussed in the preceding chapter, E. Geoffroy Saint-Hilaire later derived the concept of homology from the idea of a *unité du type*, which in turn was rooted in the *échelle des êtres*, a dominant concept of 18th century biology. And it is from the *échelle des êtres* that the model of recapitulation by terminal addition, as proposed by Meckel and Serres, is to be deduced (Gould, 1977).

Discussing the nervous system of invertebrates, E. Serres referred to his earlier studies on the central nervous system of vertebrates which, in his own view, had shown that "the embryos of higher vertebrates successively recapitulate the permanent forms of lower vertebrates; that therefore all dissimilarities in the structure of the cerebro-spinal axis have been produced by some metamorphoses more or less" (Serres, 1824: 378). As metamorphoses are added to the end of an existing ontogenetic sequence, development proceeds along the ladder of life. It is at the same time the concept of metamorphosis which allows to reduce the multiplicity of organic appearances to the unity of type inherent in the Great Chain of Being (Serres, 1827a: 51-52). It has been discussed in the previous chapter how Serres advocated the decomposition of organic structure into its constituent elements in order to render a comparison in terms of "analogy" (homology in the modern sense) possible; his discussion of the development of the kidneys was cited as evidence of the atomistic background of his view of life: "In the theory of formations, and also in nature, an organ is almost always a composite structure, a combination of several parts of the organ, or of several layers superimposed one on the other..." (Serres, 1827a: 60) - as in the case of teeth. The atomistic conception of organization renders terminal addition to an existing type easily explained: parts may either be added, or subdivided, or elements may coalesce.

According to Serres, an increase in complexity of organization results either from the addition of parts, or from additional metamorphoses of constituent elements of the existing structure. The idea of additional steps of metamorphosis transcends the orthodox atomistic conception of life and opens the field of vision to include aspects of the epigenetic model of ontogeny and therewith the concept of deviation. Orthodox atomism would predict a linear order of life, the sequence resulting from a progressive addition of elements to the existing form. Metamorphosis of constituent elements representing a given type of organization may result in deviation from an existing ontogenetic pathway, and therewith in a branching order of nature (see below). Serres (1827b: 108) came close to the epigenetic viewpoint in his discussion of E. Geoffroy Saint-Hilaire's attempt to homologize the bones in the head of man and of fishes. Finding that the elements are more numerous in the fish skull, Geoffroy tried to identify the "analogues" of the fish bones in comparison to the ossification centers observed in the embryo human skull. In contrast to the more frequently invoked growth by apposition or juxtaposition, Serres came up with a theory of "*fractionnement*", finding that the number of the cranial elements in the human skull becomes progessively increased by subdivision as one descends to the level of fishes; the heart, however, shows the opposite tendency, becoming progressively fractionated as one ascends from fishes to mammals (Serres, 1827: 121). The corollary of the epigenetic theroy of *"fractionnement"* is that development always proceeds from the simple to the more complex: "...a more complicated form is always preceded [during ontogeny] by a more simple structure..." (Serres, 1827b: 82).

Metamorphosis of some constituent element, for example of the heart, by means of compartmentalization or *"fractionnement"* might be viewed to result in a branching sequence of differentiation. Investigating the development of the brain of higher mammals, and having recognized the correspondence (homology) of its constituent parts throughout the vertebrates, Serres found it necessary to explain the differences in brain structure between the vertebrate classes: "...one had to trace all its [i.e., the brain's] metamorphoses in all the classes..." (Serres, 1827b:138). At this point, Serres might have gone on to elaborate on the *divergence* of the brain in the different vertebrate classes; instead, he stuck to the linear *échelle des êtres*: the brain of higher mammals was described to go through a "series of successive metamorphoses" (Serres, 1827b: 142), thus recapitulating stages typical for fishes, reptiles and birds in succession.

The reason why Serres continued to adhere to a linear arrangement of organisms may be found in the explanation he offered accounting for the mechanism driving this series of metamorphoses. He invoked a *force formatrice* (Serres, 1830: 48), a formative force which recalls the *Reproduktionskraft* postulated by Kielmeyer (1793). The reproductive energy was believed to transmit some impetus to the ontogenetic process. If the effect of the *force formatrice* was superabundant, new terminal stages of metamorphosis of the constituent elements would be added to the existing ontogeny. If, however, the effect of the developmental force was deficient, ontogenesis would come to a halt at some stage reflecting the permanent structure of lower types. As Meckel (1811: 2) had done before, Serres (1830) derived the empirical evidence for the existence of the *force formatrice* from teratology. *Monstres par excès* would result from a surplus of developmental force, *monstres par défaut* on the other hand were thought to be produced by a deficit of the formative power. This is how atavisms found their natural explanation.

It was left to Ernst Haeckel to provide an evolutionary explanation for the Meckel-Serres Law of Recapitulation (Russell, 1916). As organisms higher on the scale of beings go through permanent stages characterizing lower animals during their ontogeny, they would recapitulate in an abbreviated manner their phylogenetic history. Terminal addition, resulting in an increase of complexity which in turn is used as a measure of progressive development, was easy to explain on the basis of atomistic philosophy. Darwin's theory of pangenesis, invoked to explain problems of heredity and variation, comes close to George Buffon's atomistic theory of organic molecules: "I have read Buffon: whole pages are laughably like mine", exclaimed Darwin (Fr. Darwin, 1887, vil. II: 375; see also Bowler, 1974: 175; Mayr, 1982: 694). This testifies to an atomistic background of the Darwinian theory of evolution (Rieppel, 1986a; see also chapter 2), a philosophical conception of organization also adopted by Haeckel. It has been discussed above (chapter 2) that the atomistic analogy to the formation of organisms was the growth of crystals. This analogy was of paramount importance for the rise of the cell theory as developed by Theodor Schwann (Russell, 1916: 182, 184), whose views may have influenced the argumentation of Haeckel: "There is the same talk of cells as organic crystals, of crystal trees, of the analogy between assimilation by the cell and the growth of crystals in a mother liquid" (Russell, 1916: 248). Small wonder that Haeckel, the famous investigator of radiolarians, was striving for a "crystallography of the organic" (Russell, 1916: 249). Physical forces would drive the progressive development of life to ever higher levels of complexity by adding to what has already been achieved during the foregoing historical process.

Haeckel (1902) viewed evolution as a necessary process as Darwin had proposed. He furthermore distinguished two laws of evolution, the law of adaptation and the law of progress; the latter law was considered to represent the "most general result of paleontological investigations" (Haeckel, 1902, vol. I: 275). As was necessarily implied in his theory of recapitulation by terminal addition, Haeckel found the course of progressive development reflected in the "*Stufenleiter*" (scale of beings) which consitutes the object of interest of the comparative anatomist: "Comparative anatomy explains to us the relations in respect to which the classes of vertebrates form a ladder ascending from fishes upwards through the amphibians to the mammals." (Haeckel, 1902, vol. I: 312). Since both the classification (in terms of a scale of beings) and fossils reflect the general progress of evolution, there must be a parallelism between the results of paleontology and comparative anatomy. Recapitulating phylogeny, ontogeny is the third line of evidence joining up in the threefold parallelism (Russell, 1916: 254-25) evoked in support of the law of progressive development.

Haeckel was careful, however, to avoid some of the pitfalls that might result from an uncritical adoption of the Meckel-Serres Law in the context of evolutionism. One of the problems was that by terminal addition, the ontogeny of descendant forms would become increasingly longer. Haeckel and his followers therefore had to admit various mechanisms of abbreviation of descendant ontogenies such as acceleration and/or deletion (Gould, 1977: 83). Another problem was that the Darwinian theory of evolution provides an explanation for a branching pattern of order in nature (Ospovat, 1981), as Darwin himself had made clear by the inclusion, in the fourth chapter of the *Origin*, of a branching diagram in order to illustrate the mode of action of natural selection. Recapitulation by terminal addition, however, was tied to a linear hierarchy of life. To complicate matters even further, the process of recapitulation took care of the law of progressive development, but not of the law of adaptation. Haeckel therefore proceeded to distinguish the process of *palingenesis* from that of *caenogenesis*.

Palingenesis was characterized by Haeckel as the *Auszugsgeschichte* (translated as "epitomized history" by Gould, 1977: 82) of the individual organism. Palingenesis is the Greek word for rebirth. In neoplatonic philosophy, the notion came to mean rebirth at an advanced stage of illumination (Shumaker 1972: 230). Bonnet (1769) used the term with a similar connotation, designating the ascent of the species up through the Great Chain of Being by a series of rebirths from preformed "germs of resurrection". Haeckel used the term palingenesis to denote a condensed synopsis of progressive development during phylogeny. The palingenetic development of an organism recorded, in an abbreviated manner, the latter's ascent on the ladder of life.

Caenogenesis on the other hand was characterized as *Störungsgeschichte* (translated as "falsified history" by Gould, 197: 82), corresponding to all those aspects of ontogeny which would tend to blur the linear course of palingenetic development. Caenogenesis was above all introduced to account for special adaptation of larval stages, but Haeckel came to include under this heading the phenomena of heterochrony, too, i.e. the "temporal and spatial dislocations in the order of inherited events" (Gould, 1977: 82). Kluge and Strauss (1985: 252) identified caenogenesis as non-terminal addition or substitution. In short, palingenesis took care of the law of progressive development, while caenogenesis accounted for the law of adaptation (Haeckel, 1902: 311). The corollary is that only part of the developmental process reflects the scale of beings and the succession of fossils through time: palingenesis provides an "*idealized*" image of part of the phylogenetic process only, namely the progression during evolution through a series of increasingly complex types of organization. The phylogenetic process as a whole accounts for a branching order of life, as Haeckel noted (1902: 314), and therefore it must include processes other than palingenesis, i.e. adaptation.

Haeckel introduced the metaphor of the tree as illustration of the phylogenetic process. Palingenesis provided the stem of the tree the ascent through a series of increasingly complex levels of organization. Sitting out on the branches of the tree would be the real organisms, representing a given level of organization, but at the same time adapted to their specific environment (Rieppel, 1985c).

Etienne Geoffroy Saint-Hilaire and the Concept of Deviation

Etienne Geoffroy Saint-Hilaire, friend and collaborator of E. Serres, was yet another author to proclaim that "Frédérick Meckel was the first to enter these new paths to cognition, giving close attention to the successive formations and to the mechanics of development of the embryo and foetus...he realized quickly that if the perfect being develops according to uniform laws, the imperfect being would naturally result from a deviation of these same laws...Frédérick Meckel interpreted malformations as a retardation of development; retardation of one part only, as the rest of the organism continues to develop in a regular manner" (Geoffroy Saint-Hilaire, 1825b: 84). His introduction to a report of E. Serres' work proves of interest in several ways. Like Meckel and Serres, Geoffroy Saint-Hilaire used teratological investigations to substantiate his theory of recapitulation. As Meckel and Serres, Geoffroy Saint-Hilaire postulated recapitulation not with respect to the entire organism, but of organs or organ-complexes, i.e. of constituent elements of the organism only. In contrast to Meckel and Serres, however, Geoffroy Saint-Hilaire tended to emphasize *deviation* rather than terminal addition, which would result in a branching order of nature rather than in a linear series of forms.

Still, Geoffroy Saint-Hilaire's theory of recapitulation would appear again to have been initially related to the concept of the *échelle des êtres*: "...try to discover, with a mind freed of the illusions and prejudices of the past, the succession of differential facts in the *évolution* [ontogenetic development] of an organism which has passed through all the stages of its life, and you will find, in a condensed form, the spectacle of the evolution of the terrestrial globe, that is to say a succession of differential facts one giving rise to the other...we observe the transformation of the organization while passing from the condition of one class of animals to that of another class: such, an amphibian is first a fish, named as a tadpole, and then becomes a reptile under the name of a frog" (Geoffroy Saint-Hilaire, 1833a: 81-82; note that in the classifications of the time, the amphibians were frequently included within the class of the reptiles). The amphibian passes from fish to reptile, as it does in the serial arrangement of the scale of beings. This would seem to imply recapitulation by terminal addition, rooted in the atomistic tradition, but the quotation given discloses Geoffroy's adherence to epigenesis: evolution - be it ontogenetic or transspecific - consists of a succession of different stages, the succeeding one produced or created by the preceding one, as the principle of continuity demands.

The principle of continuity dictates the harmonious correlation of the parts of an organism, due to the developmental and functional integration of the constituent elements of organic structure. To deny the possibility of the derivation of vertebrates from cephalopods amounts to the admission that organs, "which can only exist if produced one by the other because of the reciprocal interactions of the nervous and circulatory systems between them, would refuse to belong to each other, would refuse to be in a harmonious relation to each other; such a hypothesis is inadmissible, however: as soon as harmony ceases to exist between the organs, life comes to an end" (Geoffroy Saint-Hilaire, 1830: 48). Remember that Geoffroy Saint-Hilaire was the founder of the *principe des connexions*, a methodological tool to trace the *unité du type* as discussed in the preceding chapter. He assumed the existence of a "primitive plan of organization" composed of a fixed number of constituent elements which stand in a constant relative position to each other. Organic appearances represent the manifold variations of this common structural plan. Here he gives us the reasons for the necessity of such a common structural plan, permitting the reduction of vertebrates and cephalopods to the same basic type of organization: number and relative position of "analogous" parts (homologous parts in the modern terminology) is determined by their epigenetic development and functional integration, the latter following particularly from innervation and circulation. Developmental and functional integration determine the harmonious relations between the constituent elements of a given ground-plan of organization: ...all parts of an animal must coincide with each other, so that each may receive the principle of unity of action and movement" (1833b: 44-45).

The *principe des connexions* depends on the constant number and relative position of the constituent elements of a structural ground-plan as follows from the developmental and functional integration of these parts. The unity of type is preserved as the structural ground-plan is modified in adaptation to environmental conditions, because the developmental and functional ties relating the constituent elements with each other cannot be disrupted without loss of the harmony of organization - which would mean death. Teratology provided examples, according to Geoffroy Saint-Hilaire, for the fact that modifications of blood circulation by an increase or decrease of the diameter of the vessels may change the volume of certain organs, enlarging some and reducing others (Geoffroy Saint-Hilaire, 1825e: 241-242). In a similar vein it could be argued that environmental conditions led to an increase in volume and change of function of the auditory ossicles as seen in mammals, resulting in the opercular series of bones serving respiration in water (see chapter 3) with no disturbance of the harmony characterizing the vertebrate type (Geoffroy Saint-Hilaire, 1825a).

If adaptational modifications must not disrupt the harmony of a "*fond commun d'organisation*" (Geoffroy Saint-Hilaire, 1833a: 66), it follows that transformation remains restricted to the limits of a given type: "Every organ system is...in all organisms restricted to some limits of variation..." (Geoffroy Saint-Hilaire, 1828a: 213). The limits of variation are set by the "*Loi du Balancement des Organes*" ("law of equilibrium between organs": Geoffroy Saint-Hilaire, 1825c). This is a concept first developed by Kielmeyer (1793), stating that some part must be reduced if another is increased in order to preserve the equilibrium and therewith the harmony of organization. Geoffroy Saint-Hilaire illustrated this law by a discussion of the palate of crocodiles in comparison to that of lizards. Crocodiles are characterized by a secondary palate, developed by medial palatal flanges of the premaxilary, maxillary and palatine bones in mesosuchians (to which the teleosaurs described by Geoffroy Saint-Hilaire belong), and including the pterygoid bones in eusuchians, i.e. in modern type crocodilians. Lizards lack a secondary palate, however: premaxillae, maxillae (and palatines) do not meet along the ventral midline of the skull. Geoffroy Saint-Hilaire claimed that the maxillary (and palatine) bone of lizards, in particular of varanids, is relatively longer as compared to crocodiles, but :"...they can increase their volume in that [viz. longitudinal] direction only if losing substance in their width" (Geoffroy Saint-Hilaire, 1825d: 131). It was mentioned that an increase in diameter of a blood vessel may increase the blood supply to a given organ or organ system which, as a consequence of this stimulus, would grow to larger size. Given an absolute blood volume, the increased supply to some part of the body must result in a decreased supply to some other constituent element: the enlargement of one organ necessitates the reduction of some other. Thence follows the "*Loi du Balancement*".

Under environmental influences "the most disparate forms originate from the same foundations of organization" (Geoffroy Saint-Hilaire, 1833a: 71), but at the same time the ground-plan of organization tends to be preserved. The development of an organism is thus subject to the influence of two antagonistic forces (Geoffroy Saint-Hilaire, 1833a: 68). On the one hand, changing environmental conditions tend to impose adaptive modifications, the extent of which is, on the other hand, opposed by forces tending to preserve the "*fond commun d'organisation*" (Geoffroy Saint-Hilaire, 1833a: 66, 69). The "*plan primitif*" (Geoffroy Saint-Hilaire, 1824a: 423), "*type général*" (Geoffroy Saint-Hilaire, 1833a: 67) or "*type commun*" (Geoffroy Saint-Hilaire, 1828a: 211) tends to be preserved by the *nisus formativus*, a developmental force first proposed by Johann Friedrich Blumenbach (1781). The vertebrate type has become profoundly modified in the course of earth history, as was admitted by Geoffroy Saint-Hilaire (1828b), and to explain these modifications one must assume environmental perturbations in the past which were more dramatic than those observed today, but still the unity of the vertebrate type was preserved: ...admit the continuous action of the same *nisus formativus* [in all vertebrates], that is to say the same control of formation to produce the *fond organique* of the vertebrate animals..." (Geoffroy Saint-Hilaire, 1828b: 207). It may be noted at this juncture that the "control of formation" evoked by Geoffroy Saint-Hilaire needs not necessarily be related to the idealistic concept of the *nisus formativus*, but may well be equated with mechanistic principles of morphogenesis tending to preserve a certain *bauplan*, as was argued by Shubin and Alberch (1986). It thus becomes evident that the views of Geoffroy Saint-Hilaire approach modern structuralism, as will be more fully discussed below.

According to Geoffroy, then, developmental controls restrict variation and adaptive modification to the limits of a given type or ground-plan of organization. A corollary of this view must be that, in contrast to an insensibly graded series of forms, the various types of organization remain separated by morphological gaps. This is the point where the deviation of developmental processes comes in, replacing the concept of terminal addition. If an organism is conceived of as being composed of interchangeable parts or "atoms", variation is fundamental, potentially affecting the organism in all its parts (see chapter 2); if in addition, progressive evolution is believed to proceed by the terminal addition of new parts or new steps of metamorphosis to ancestral ontogenies, then all possible intermediate stages obtain, at least potentially, between any two levels of organization. Continuity of transformation renders the delineation of static types or common plans of organization arbitrary; there is no need nor any basis for the assumption of developmental constraints.

If, however, different types of organization are separated by morphological gaps; if the harmonious interrelation of constituent elements, prerequisite for life, is restricted to certain ground-plans of organization separated from each other by "forbidden morphologies" (Stock and Bryant, 1981: Holder, 1983), then a parallelism of ontogenetic development and evolutionary transformation can no longer be based on the atomistic model of terminal addition. Instead, deviation in the course of an epigenetic process of development must be assumed. As one organ formed produces the next one to develop, changes affecting one stage must have an effect on all succeeding stages of development: this is what must be expected on the basis of the principle of continuity. Assuming that early developmental stages undergo some modifications, these may result in a radical change of adult morphology. As the succeeding developmental stages are developmentally and functionally interrelated, one must assume that if development ends up by producing a viable adult organism, the harmonious relations between its constituent parts have been preserved, or new such relations have developed. Thus may be created, by deviation from an ancestral ontogeny, a new ground-plan of organization, separated from the ancestral type by a morphological gap.

> "It is obvious that the inferior types of oviparous animals have not given rise to the superior degree of organization, that is to the birds, by insensible transformations. Instead a possible accident, of trivial origin but of unpredictable effects, was enough (an accident affecting some reptile which I feel incompetent even to attempt to characterize) to develop in all

parts of the body the conditions characteristic of the ornithological type" (Geoffroy Saint-Hilaire, 1833a: 80).

It is the earliest stages of development of the reptile lung which, according to Geoffroy, would be affected by deviation, providing the starting point for the development of the ornithological type of that organ.

Karl Ernst von Baer and the Laws of Individual Development

The concept of deviation was fully developed by Karl Ernst von Baer in his influential treatise on the *Entwickelungsgeschichte der Thiere*, published in 1828. His embryological investigations not only resulted in the rejection of the doctrine of recapitulation by terminal addition, but they also led him to the recognition of four basic types of animal organization - quite independently from Georges Cuvier, as von Baer emphasized (1828: vii). Cuvier, in his *Règne Animal* of 1817, had classified the animal kingdom on the basis of comparative anatomy, proposing four basic groups or *embranchements* which in his view could not be reduced to one another. This classification provided the starting point for the great debate of 1830, triggered by Geoffroy Saint-Hilaire's support in favor of a paper reducing the vertebrates and cephalopods to a common ground-plan of organization. The comparative embryological work of von Baer had established the same *embranchements* as supported by Cuvier, demonstrating these to correspond to four basically different types of developmental differentiation. The corollary of such a classification, both for Cuvier and von Baer, is the rejection of the concept of the *échelle des êtres*. If there is any gradation of complexity in animal organization, it can only be established within the limits of the type, bearing on the arrangement of less inclusive groups within the *embranchements*.

The refutation of the Great Chain of Being must imply the rejection of the theory of recapitulation by terminal addition, the two concepts being closely related to each other as discussed above. Von Baer formulated a number of objections against the idea that the ontogenetic development of the individual organism should recapitulate the scale of beings such that "the higher organism passes through the permanent stage of lower organisms during its individual development" (1828: 199). His main arguments (von Baer, 1828: 202-206) derived from special larval or embryonic adaptations and heterochronic displacements of the successional order of embryonic stages, phenomena subsumed under the heading of caenogenesis by Haeckel (see above). There was, however, an additional aspect to account for, namely the adaptation of any living being to its particular environmental conditions. The embryo of a tetrapod could never pass through the adult stage of "a fish", because such a "fish" does not exist; rather, there exists a multitude of fish species, each adapted to a particular mode of life and therefore too specialized to represent a transient stage in the ontogeny of a tetrapod (von Baer, 1828: 231). This argument will have to be more fully developed below.

The theory of recapitulation by terminal addition was (and is) not only based on the *échelle des êtres*, it was (and is) also linked to an atomistic background, at least in the form advocated by Meckel, Serres and Haeckel. The Aristotelian von Baer (1828) criticized the atomistic conception of organization, since his work in embryology supported an epigenetic type of development. Epigenesis was characterized above as a process of growth (budding) and compartmentalization (segmentation, differentiation). Von Baer noted that the formation of organs results from a process

"of continuous separation, with the difference, however, that the organs do not become detached from each other, because they never become a whole, but always remain parts. The theory of Serres is radically opposed to this view. Following this author the whole organism would be formed by the   fusion of originally separated rudiments, so that even the most simple elements would be composed of two halves..." (1828: 157).

What von Baer meant by "separation without detachment" corresponds to a process of epigenetic differentiation. Development starts off from a homogeneous primordium and leads up to a heterogeneous adult condition through a continuous process of growth, compartmentalization and individualization or differentiation of the compartments. Von Baer distinguished three different types of differentiation: the "primary differentiation" corresponds to the germ layer formation; "histological differentiation" takes place within these germ layers; "morphological differentiation" accounts for the formation of organ systems (Russell, 1916: 115). Through the process of differentiation, "the homogeneous and generalized gives rise to the heterogeneous and specialized" (von Baer, 1828: 153).

At this point the radical difference between von Baer's concept of epigenetic deviation and the theory of recapitulation by terminal addition becomes apparent. The Meckel-Serres Law implies that organs or organ systems of higher organisms pass through developmental stages corresponding to the adult condition of the same organ or organ-system in animals of an inferior type of organization, standing lower on the scale of being. Von Baer, in comparison, showed the developmental process to proceed along pathways characteristic of four radically different types. Within these types, development starts out from a homogeneous primordium of similar structure and appearance. Von Baer even visualized a germinal primordium (a simple vesicle) as starting point for the development of all animal life. As development continues, the primordium of a given type becomes progressively differentiated and individualized: to von Baer (1828: 224), ontogenesis was a process of individuation, governed by the four laws of "individual development". The more advanced different representatives of the same basic type - say different vertebrate animals - are in the process of differentiation and individualization, the more different they will be, i.e. the more will they have diverged from one another. In other words, differentiation taking place at the very beginning of development is characteristic for all representatives of the same basic type: features appearing early during ontogeny are widely distributed within the type. As differentiation continues and deviation sets in, features appear which are characteristic of some less inclusive group only, characterizing a subtype within the basic ground-plan of organization. These features are less widely distributed than earlier ones, they are more specialized and characterize a less inclusive group.

> "When one observes the progress of the formation [of the embryo] what first of all leaps at the eye is that there is gradually taking place a transition from something homogeneous and general to something heterogeneous and special. This law of [embryonic] formation it is quite impossible to fail to recognize, and it so completely dominates every single moment of the embryo's metamorphoses that no accurate account of its development can be expressed except in the foregoing terms" (von Baer, 1828: 153, translated by Lovejoy, 1959b: 438).

The immediate corollary of this epigenetic view of differentiation are the laws of individual development as formulated by von Baer (1828: 224):

> 1) Common features of a more inclusive group of animals develop earlier in the embryo than special features.

2) From the most general characteristics of form develop the less general characteristics until finally the most specialized features make their appearance.

3) Each embryo of a given animal form, instead of passing through the definite forms of other animals, deviates from the latter.

4) Therefore, an embryo of a higher animal form never corresponds to another animal form, but only to the latter's embryo.

The latter two laws constitute a refutation of the theory of recapitulation by terminal addition (see above), and will not be considered any further in the present context. The first two laws, however, are of particular interest, since they provide the basis for the concept of deviation. Consider the development of a trout, a frog, a lizard and a chick: fishes, amphibians, reptiles and birds share similar features during early ontogenesis which are the characters diagnosing the Vertebrata. "The embryo of a vertebate is a vertebrate from the very beginning of its development..." (von Baer, 1828: 220), and cannot, therefore, be reduced to a common structural plan shared with cephalopods, as Geoffroy Saint-Hilaire would have it. Baer (1828: 221) noted that he owned "two embryos in spirit which I have forgotten to label, and now I am quite unable to determine the [vertebrate] class to which they belong...". But as development continues, the frog, the lizard and the chick will continue to share characters, while deviating from the form of the developing trout. At this intermediate stage of differentiation, the frog, the lizard and the chick will share features which are less widely distributed within the Vertebrata as opposed to the characters developed earlier; these later features therefore diagnose a less inclusive group within the Vertebrata, a subtype which can be named Tetrapoda.

The trout on the other hand undergoes its proper differentiation or individualization, leading up to its adult form adapted to its particular mode of life. During its earliest developmental stages, the fish was but an "imperfect vertebrate"; at the end of its ontogenesis, the fish is not only a vertebrate, but also a "perfected" fish (von Baer, 1828: 231). The frog, the lizard and the chick therefore could never have passed through the stage of a "perfected" trout! They do, however, share similarities with the early fish embryo, because they too, during their earliest developmental stages, are nothing but "imperfect" vertebrates. This corresponds to the distinction of the "type" from the "grade of its development" (von Baer, 1828: 207). Both fish and tetrapods belong to the same type, and this is why they resemble each other at an incomplete grade of development of that type - when they are but "imperfect" vertebrates. As the individualization of the fish progresses, it reaches a higher grade of development, a higher degree of differentiation, and finally becomes a "perfected" fish that bears no resemblance to an embryonic or an adult tetrapod: no recapitulation of adult structures obtains.

As the development of the frog, the lizard and the chick proceeds from the common stage of an "imperfect" tetrapod, the frog embryo's differentiation will diverge from that of the lizard and bird. Again, the frog will become "perfected", i.e. adult and adapted to its conditions of life, while the lizard and bird embryos continue to share similarities. These features, originating still later in development, after the divergence of amphibians, characterize a subgroup of the Tetrapoda, a subtype named Sauropsida (constituting the Amniota together with mammals). Continuing development will finally result in the divergence of the lizard and the chick, bringing about the differentiation of features characteristic of still less inclusive groups or subtypes, viz. the reptiles and birds.

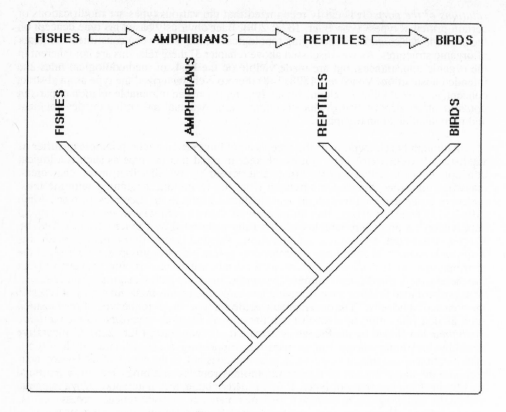

Fig. 5: Recapitulation by terminal addition is linked to a serial arrangement of organisms (above). The doctrine of deviation, in contrast, results in a branching pattern of order in nature (below).

As becomes immediately obvious, von Baer's theory of epigenetic development implies recapitulation of similar ontogenetic stages: distantly related groups belonging to the same basic type will share early ontogenetic stages only; progressively less distantly related organisms will share a progressively longer segment of their developmental pathways before deviation sets in. Von Baerian recapitulation does again parallel the results of comparative anatomy, i.e. classification, but the topography of this classification is different from that based on recapitulation by terminal addition. Terminal addition must be related to a serial arrangement of organisms, whereas the concept of deviation produces a hierarchy of groups within groups, corresponding to a branching order of life and symbolized as follows: (Vertebrata (Tetrapoda (Sauropsida (Reptilia) (Aves)))). What von Baer's first law specifies is that characters of more inclusive types develop prior to characters of less inclusive types; from this, von Baer inferred his second law stating that the more general gives rise to the more particular in the course of ontogeny.

The epigenetic process of differentiation yields a hierarchy of types and subtypes. "As *type* I designate the topographical relations of organic elements or organs to each other" (von Baer, 1828: 208), or, in other words: *"The type corresponds to the topographical relations of the parts*. It is easily recognized that the various types are modifications of certain principal types" (von Baer, 1828: 208). Baer obviously based his type concept on the *principe des connexion*, specifying topological relations between constituent elements of organic structures. As was discussed above (chapter 3) these relations are not inherent in the organic appearances, but are made visible on the basis of methodological rules and intended abstraction. Von Baer (1828: 148) therefore characterized the type as an abstract concept, an idea guiding development: the type represents an immutable relation of parts *in potentia*, an element of 'being' which becomes variably actualized during the development and individuation of an organism.

Although his conception of the type as a guiding, i.e. energetic principle qualifies as vitalistic (Russell, 1916: 115), it must be kept in mind that the type as such is a logical concept. Groups within groups are characterized by the distribution of characters, classification proceeding by subdivision from top to bottom, beginning with the most inclusive group and ending with the least inclusive group in the hierarchy of types. Mayr (1982: 158) characterized this procedure as "downward classification by logical subdivision", a procedure which was first fully elaborated by Cuvier (Rieppel, 1987b). Cuvier classified by logical subdivision, starting from the top, i.e. with his *embranchements*, and continuing to progressively less inclusive groups on the basis of the distribution of adult characteristics (for further details see chapter 6). Von Baer showed that the more widely distributed features, characterizing more inclusive groups, precede less widely distributed features, characterizing less inclusive groups in the hierarchy of types, in the course of ontogeny. "The development of the embryo relates to the type of organization such as if it [the embryo] went through the animal kingdom according to the *méthode analytique* developed by the French systematists..." (von Baer, 1828: 225). Comparative anatomy and embryology run in parallel, supporting a subordinated dichotomous classification of the animal kingdom. A mammal does not, in the course of its development, recapitulate stages typical of fishes, amphibians reptiles and birds. Rather, a mammal recapitulates the hierarchy of types: it first is a vertebrate, then a tetrapod, then a mammal and finally individualizes to become a particular mammal. The ontogenetic process can thus be viewed as a recapitulation of the hierarchy of types, but this succession of levels of organization does not imply transitions from old to new, but only logical relations. As Lovejoy (1959b: 444) has emphasized, von Baer did not imply any transition from an older to a later type in any phylogenetic sense, but rather "a transition from a *logically* more 'general' and indefinite type to a logically more determinate and specific type".

Ontogeny and Hierarchy

The parallelism of ontogeny and cosmology, of the microcosm and macrocosm, is an ancient concept rooted in neoplatonic or hermetic philosophy. The temporalization of biology in the 18th century (Lovejoy, 1936) assigned to this parallelism a more precise meaning, i.e. that of a recapitulation of the Great Chain of Being: the ontogeny of a higher organism mirrors the latter's ascent through the *scala naturae*. Further refinement during the early 19th century led to the formulation of the theory of recapitulation by terminal addition as expounded by Meckel and Serres. Recapitulation was no longer meant to affect the whole organism, but related to the development of its constituent element, i.e. to its organs or organ systems. This reveals an atomistic philosophy, explicitly accepted by Serres in his discussion of the development of the kidney and the teeth, and by Haeckel in his "crystallography of the organic". Atomism implies the view of an organism as being composed of interchangeable parts; variation is fundamental, i.e. it may affect the organism in all aspects of its structure. As progressive development proceeds along the *échelle des êtres*, it must be reflected in ontogeny, or may even be thought to be powered by the mechanisms of ontogenesis: new parts, or additional steps of metamorphosis of parts already present, are added to the end of the existing (ancestral) ontogenetic sequence. An increase of complexity is thus achieved by the addition of parts or of steps of metamorphosis to the level of organization that has already been arrived at: progressive development is a cumulative process, building on the past history of a lineage. As variation is fundamental, all possible intermediates between various grades of complexity may at least potentially exist.

If ontogeny is to reflect evolutionary progression, or is viewed as the mechanism of progressive development functioning by terminal addition, the resulting order of life must necessarily be linear, as is the scale of beings. Such a linear hierarchy may be characterized as being exclusive, or in other words: the linear hierarchy (serial arrangement) characterizes exclusive taxa only. In the *scala naturae*, "each level of perfection was considered an advance (or degradation) from the next lower (or higher) level in the hierarchy, but did *not* include it" (Mayr, 1982: 206). With reference to Fig. 5 it may be stated that 'reptiles' follow 'amphibians', but are not included within the latter. To put it the other way around: the 'reptiles' is an exclusive taxon, comprising only the Reptilia but excluding everything else. By the same token, the taxa included in a *scala naturae* are privative with the exception of the terminal one: 'reptiles' differ from 'amphibians' by the addition of new features, or conversely: 'amphibians' differ from 'reptiles' by the absence of the features diagnosing the latter group.

The situation changes with the introduction of the concept of deviation, based on an epigenetic model of differentiation. Development is a process of growth, compartmentalization and individualization, each organ formed being the material cause of the next one to develop. Recapitulation does not imply the repetition of adult stages of lower (ancestral) grades of organization; rather it is claimed that the ontogeny of an organism of some higher grade of organization shares similar ontogenetic stages with lower organisms. Related organisms share developmental pathways and hence ontogenetic stages, and the closer the organisms are related, the longer will their ontogenies run in parallel before deviation sets in. Deviation of developmental pathways from some embryonic stage results in the individuation of terminal taxa in a succession of branching events. Deviation may affect ontogeny at any stage, creating new morphologies from shared similar rudiments. The earlier the stage at which deviation sets in, the more will the new morphology differ from the existing one. The epigenetic process of differentiation results in the development of more widely distributed features at an earlier ontogenetic stage as compared to the development of less widely distributed features which arise only later during ontogeny. Ontogeny leads from the homogeneous to the heterogeneous, from the more general to the less general. A hierarchy of types obtains, of groups within groups. If anything, ontogeny mirrors this hierarchy of types, not adult stages of lower grades of organization. In fact, the ontogenetic process of differentiation is a mixture of recapitulation (of the hierarchy of types) and individuation (of specific forms of life). Living beings sit out on the tips of a branching diagram which by the subordination of dichotomies specifies the hierarchy of types.

Deviation produces a branching order of life, the successive dichotomies corresponding to the hierarchy of types being recapitulated during ontogeny, or in other words, to the succession of recapitulated events of deviation. The crucial difference from the serial arrangement of organisms is that the subordinated hierarchy of groups under groups, of types within types is inclusive, or in other words: the von Baerian notion of types specifies inclusive taxa. With reference to Fig. 5 it will be noted that the lizard is a reptile, but also a tetrapod and a sauropsid. Or, to put it the other way around, the Vertebrata include fishes, amphibians, reptiles and birds, the Tetrapoda include amphibians, reptiles and birds, and the Sauropsida include reptiles and birds. The hierarchy of progressively less inclusive groups within groups, characterized by progressively less widely distributed features, corresponds to the von Baerian hierarchy of subtypes within types. Living beings represent individualized differentiations within these types and subtypes.

The crucial point to note is that the serial and subordinated hierarchies are complementary to each other. Nelson and Platnick (1981, 1984) have shown that any linear hierarchy may be translated into a subordinated one and vice versa, and they illustrate this with reference to the serial classification proposed by Aristotle (Nelson and Platnick,1981: 68-72). His simple hierarchy of souls may serve as an example in the present context (Fig. 6). It will be recalled that Aristotle was the founder of the principle of continuity, and that his hierarchy of souls laid the basis for the concept of the *scala naturae*. It has also been shown that his writings can be interpreted so as to imply terminal addition. Plants are characterized by the vegetative soul only, animals add the sensitive soul during later stages of ontogeny, and man becomes endowed with the rational soul during postembryonic development. There is a perfect parallelism of ontogeny and the scale of beings. This linear arrangement may now be translated into a dichotomous and subordinated hierarchy (as exemplified in Fig. 6) by the specification of inclusive taxa: the taxon characterized by the vegetative soul includes plants, animals and man, The sensitive soul diagnoses a taxon including both animals and man.

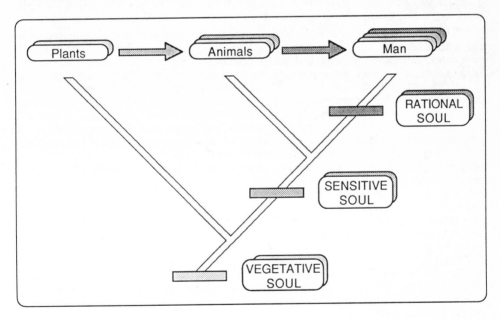

Fig. 6. The translation of the Aristotelian linear hierarchy of souls, recapitulated during ontogeny, into a dichotomous subordinated hierarchy by the specification of inclusive taxa.

The vegetative soul characterizes a group, 'living beings', which includes plants, animals and man. The sensitive soul, originating later during ontogeny, is less widely distributed within the group of 'living beings' as compared to the vegetative soul, and it therefore characterizes a subgroup only, the animals, included within the 'living beings' and itself including man. The rational soul becomes actualized even later during ontogeny, and consequently is of even more restricted distribution, characterizing a subgroup of the animals, viz. man. It may be noted in this context that the classification of man with animals is a necessary and logical consequence of the specification of inclusive taxa. Celebrated as a heroic act of Linnaeus opposing clerical orthodoxy, his grouping of man with animals logically followed from his subordinated scheme of classification of groups within groups!

The possibility to translate the Aristotelian *scala naturae* into a subordinated hierarchy of inclusive taxa results from the fact that his diagnosis of the exclusive taxa ('plants', 'animals') is privative, i.e. based on the absence of characters. Plants differ from aninmals by the absence of the sensitive soul, animals differ from man by the absence of the rational soul.

The privative diagnosis of the exclusive taxa is in accordance with the concept of terminal addition. Indeed, the Aristotelian classification is a classical example for terminal addition: the sequence of taxa along the *scala naturae* is defined by the successive addition of new qualities (or characters) to the exclusion of all simpler stages of differentiation. Animals become diagnosed by the addition of a quality which is absent in plants; to put it in other words: the absence of sensitivity excludes plants from the animal kingdom. Only the terminal taxon in the series is positively defined: man is the only living being endowed with a rational soul.

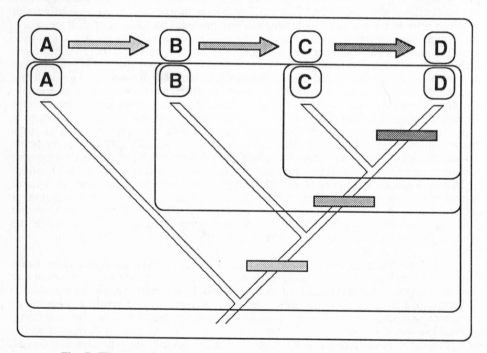

Fig. 7: The complementarity of the serial and dichotomous subordinated hierarchy.

The translation of this linear series of forms into a subordinated hierarchy of inclusive taxa shows that the latter is again specified by the same successive addition of new qualities (or characters). Palingenesis, originally thought to result from terminal addition, is now transformed into the axis (lower edge) of a branching diagram which specifies the subordinated hierarchy of types (inclusive taxa). Or, to put it in other words: the axis of a branching diagram specifies a hierarchy of types and subtypes by a succession of additive steps of differentiation. Movement from top to bottom along the lower edge (right-hand side) of the branching diagram (i.e. along the axis of the cladogram) discloses a succession of ever more inclusive predicates: man is not only an animal, but also a living being. Conversely, as one moves from bottom to top along the hierarchy of subordination, one moves to ever less inclusive levels characterized by the addition of ever more restrictive predicates: all men are living beings, but not every living being is also a man. In other words, a linear hierarchy resulting from terminal addition translates into a serial sequence of dichotomies: Riedl (1975: 161, Fig. V11-13) used the term "sequential hierarchy" to denote the topography of such a hierarchy, reproduced schematically in Fig. 7 above. Each more inclusive level includes all less inclusive groups.

As long as von Baer's first law of individual development holds, it specifies a hierarchy which is complementary to the series of privative taxa obtained from terminal addition. The law predicts the sequential development of the homologies characterizing subordinated groups of organisms as one moves up along the lower right-hand side (axis) of the cladogram (Patterson, 1983). Following the pattern of terminal addition, a sensitive soul is added to a vegetative soul, followed by the addition of a rational soul. The *scala naturae* obtains. Von Baer's first law predicts that, in any organism, the vegetative soul develops prior to the sensitive soul, which in turn precedes the rational soul. If the soul is the feature characterizing 'living beings', the successive development of a vegetative, sensitive and rational soul specifies a sequence of successively less inclusive subtypes within the type of 'living beings'. At the same time it is obvious that the vegetative soul must be more widely distributed than the rational soul within 'living beings'. This is all that is specified by the first law! But there is also a basic difference between terminal addition and von Baerian differentiation. Terminal addition states that every living being passes through the stage of a plant during it embryonic development. Some living beings remain at that stage, while others continue to develop into an animal stage, and so on. This is the linear sequence. In contrast, von Baerian recapitulation would imply that every being passes through a "plant-like" stage (the plant type) during development. As development continues, some living beings become fully differentiated plants, i.e. they develop qualities (or features) which characterize plants. Plants are no longer typified by the mere absence of animal characteristics. They become *individualized*, to use von Baer's (1828) wording, as specific plants, as a specific form of life, with the consequence that their development deviates from all those living beings which, during their ontogeny, actualize the "animal-like" stage, and so on. There results a subordinated or branching hierarchy.

The same argument can be made referring back to the trout, the frog and the lizard (Fig. 5). In von Baer's view, the trout does not represent an imperfect frog which in turn does not represent an imperfect lizard. A trout cannot be turned into a frog by the addition of limbs, just as a frog cannot be turned into a lizard by the terminal addition of metamorphotic steps to its ontogeny. Instead, the trout represents an individualized being, adapted to a specific mode of life, the differentiation of which necessarily deviates from that of a tetrapod at an early ontogenetic stage. The latter specifies the type to which they all belong: the Vertebrata.

In von Baer's terms, each branch in the subordinated hierarchy of types is individualized by the development of its own specific (or individual) qualities (or traits). The terminal taxa no longer correspond to a linear sequence of privative groups. The *scala naturae* becomes converted into a sequence of abstract types defining the axis of a branching diagram. The subordinated hierarchy of types results from the sequential differentiation of new features according to von Baer's first law of individual development. The individuation of terminal taxa deviates, i.e. branches off from this axis, from this sequence of typical stages of differentiation.

It follows from the above that the Aristotelian concept of a *scala naturae*, composed of privative taxa (except for the terminal one) and based on the concept of terminal addition, is directly compatible with the principle of continuity and hence with the notion of continuous transformation, whereas the von Baerian hierarchy is not. In the *scala naturae*, the relation of ancestry and descent is one of simple addition to the stage of differentiation already arrived at. The von Baerian hierarchy, in contrast, predicts that the vertebrate features develop before the tetrapod features, but this only specifies that the tetrapod ancestor must have been a vertebrate animal. In other words: the relation of ancestry does not permit the individuation of ancestral taxa (see chapter 6). No type or subtype can be transformed into one another: ancestry becomes a problem of deviation of the ontogenetic process of differentiation, a continuity of causes thus giving rise to a discontinuity of appearances!

This corollary of von Baer's conceptualization of ontogenetic development is emphasized by his second law of individual development. The latter makes a different prediction: it does not specify the sequential development of "typical characteristics" within whole organisms, i.e. the sequential recapitulation of the hierarchy of types during the process of individualization. Instead it addresses structural relations in character differentiation (Patterson, 1983: 25). The more universal condition of form precedes the more particular, the more general giving rise to the more specialized. The second law allows not only for deviation in the sense of non-terminal insertions, but also for bifurcations which do not specify a less inclusive group within a more inclusive level, but which oppose groups of equal hierarchical rank to each other. In other words: deviation in the sense of bifurcation may introduce a branching point into the serial arrangement of organisms. "Is it possible that branching points are involved in the ascent of the scale of beings?", asked Charles Bonnet (1764, vol. I: 59), and envisaged the possibility that the frog and the lizard represent two separate branches coming out of the insects. Should this be so, the lizard could no longer follow the frog in a linear sequence of forms; instead, the two would be opposed to each other as a consequence of the bifurcation of the linear sequence. In other words, a bifurcation of developmental pathways produces not a sequential, but an alternating subordination of dichotomies. The *scala naturae* no longer corresponds to the axis of what Riedl (1975) termed a sequential hierarchy, it no longer corresponds to the lower edge of what Panchen (1982) designated as a "Hennigean comb". Instead, the groups becomes opposed to each other in an alternating subordination of dichotomies.

If deviation is to designate non-terminal addition rather than bifurcation, it will not alter the shape of the hierarchy which remains complementary to a linear sequence, but it will alter the criteria of its reconstruction. It is no longer the sequential development of characters of groups within groups which specifies the hierarchy of types, but the relative distribution of characters within a more inclusive group: the more universal condition of form gives rise to the more particular. If, within tetrapods, the development of an amnion is inserted at a non-terminal position into an existing (ancestral) ontogeny, it defines a less inclusive group within the Tetrapoda, viz. the Amniota, in spite of the fact that tetrapod characteristics develop only after the formation of the amnion. Translation of this hierarchy into a serial arrangement does not imply that the Amniota may not follow the anamniote tetrapods (Amphibia) in sequence. The right-hand lower side (axis) of the cladogram continues to specify the differentiation of types within types, but not by the sequential development of the characters typifying the subordinated levels of the hierarchy, but by the sequence of *epigenetic* structural relations: the amnion develops as an outgrowth of more widely distributed structures, tissues or primordia (Patterson, 1983: 25). The first law of von Baer thus turns out to be a special case of his second law. The second law specifies generality of structural relations which may (or may not) develop in sequence.

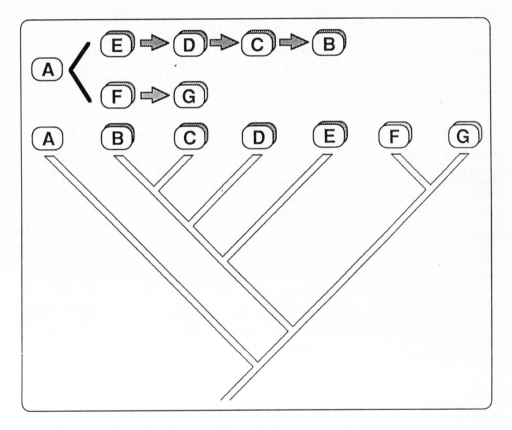

Fig. 8: Bifurcation of developmental pathways produces two groups of equal hierarchical rank being opposed to each other, [BCDE] representing the sistergroup of [FG]. These two groups cannot follow in sequence in a linear classification, bifurcation producing a branching point in the series of forms.

Deviation in the sense of bifurcation will produce a branching point in the serial arrangement of forms, however. Development may start out from a common embryonic rudiment, producing from it, as a result of a developmental bifurcation (Oster and Alberch, 1982), alternative characters diagnosing groups being opposed to each other. The early stages of differentiation of hair and feathers are similar, but their subsequent development involves a developmental bifurcation: the epidermal placode evaginates in the formation of feathers, whereas it invaginates during the initial development of hair (Oster and Alberch, 1982). Mammals and birds, characterized by hair and feathers respectively, cannot follow each other in sequence, but must be opposed to each other as groups of equal rank (Gardiner, 1982): a branching point is introduced into the linear scale of vertebrate organization. The relation of terminal or non-terminal addition on the one hand, and bifurcation on the other, and the complementary hierarchies of serial versus subordinated arrangement, are graphically depicted in Fig. 8. It becomes clear that a hierarchy resulting from a succession of bifurcations, starting out from similar early ontogenetic rudiments and producing strictly alternative character states with no terminal or non-terminal additions involved will show a topography of a strictly alternating subordination of dichotomies (Riedl, 1975: 161, Fig. V11-11). This corresponds to a serial arrangement branching at each step, i.e. the subordinated and serial arrangement will have the same topography. The normal case to be expected is some mixture of the two types of processes as outlined above (Fig. 8).

If a linear hierarchy can be translated into a subordinated one and vice versa, if the *scala naturae* can be equated with the axis of a sequential hierarchy of types (with the lower edge of a "Hennigean comb": Panchen, 1982) it must follow that terminal addition (Haeckelian recapitulation or palingenesis) is but a special case of deviation (non-terminal addition and/or bifurcation) resulting in von Baerian recapitulation (Lovtrup, 1978). Haeckelian recapitulation by terminal addition means either the addition of new parts, or the addition of successive steps of metamorphosis to the end of an existing (ancestral) ontogeny. Deviation or von Baerian recapitulation means the insertion of new parts or of new steps of metamorphosis in a non-terminal position (sequence) of an existing (ancestral) ontogeny. This is the reason why acceptance of von Baer's model of epigenetic development does not necessarily refute Haeckel's view of recapitulation as a whole. Haeckelian recapitulation is no law, but it may occur - as a special case of von Baerian recapitulation. Commenting on the use of teratologies as evidence for recapitulation by terminal addition, von Baer (1828: 232) himself made clear that monstrosities do not represent the adult stage of some lower animal. Rather, they originate by retardation of some of their parts which stop development at an early stage of differentiation. "Sometimes, however, a similarity with some other animal in some parts is obvious, however" (von Baer, 1828: 232). These exceptional cases result from paedomorphosis in animals which have deviated from other forms by terminal addition.

If Haeckelian recapitulation is but a special case of von Baerian recapitulation, it must follow that Haeckelian palingenesis recapitulates the von Baerian hierarchy of types. As discussed above, Haeckel defined palingenesis as a condensed synopsis of progressive evolution along an idealized scale of beings. Von Baer criticized the idea of the *scala naturae*, because his view of development would produce a branching order in nature. However, he accepted a gradation of levels of increasingly complex organization within his four basic types, as is reflected by the hierarchy of types and subtypes. During his ontogeny, man would first pass through the stage of an "imperfect vertebrate", followed by the differentiation of tetrapod characteristics, mammalian characteristics originating still later, and only during postembryonic development would the infant develop into a rational being. The ontogenesis of man thus recapitulates his ascent through a hierarchy of increasingly complex types of organization corresponding to an abstract version of the scale of beings. It differs from the original concept of the *scala naturae* in that it is not the sequence of exclusive taxa which is recapitulated, but rather a sequence of inclusive types of organization. Man never passes through a fish stage, an amphibian stage or a reptile stage, but the fish deviates less from the vertebrate type than man, or, in other words, man's position is higher up on the branching diagram (cladogram) of the hierarchy of vertebrate subtypes than that of fishes. Starting from the abstract vertebrate type, more events of deviation lead up to man than are included in the differentiation of fishes. The types, i.e. the inclusive taxa, become defined at the nodes of the cladogram, the succession of these nodes corresponding to nothing else but the abstract concept of the scale of being evoked by Haeckel's palingenesis.

If the cladogram in Fig. 7 is interpreted as a phylogenetic tree, it follows that its lower edge mirrors the course of progessive evolution. The hierarchy of types in the von Baerian sense, i.e. Haeckel's palingenesis, survives as stem of the phylogenetic tree from which groups of organisms (subtypes) deviate at increasingly more complex levels of organization. As Lovtrup (1978) put it: "It is evident that the generalizations of von Baer correspond closely to a phylogenetic classification of Vertebrata, and if one accepts that the latter represents a theory on the course of vertebrate phylogeny, then von Baer's laws imply that the course of changes observed during ontogeny recapitulates the course of phylogenetic evolution". Ontogeny runs along the lower edge of the cladogram which also reflects the course of progressive evolution (Lovtrup, 1977, 1982). There arises a basic problem, however, if the hierarchical classification of groups within groups, graphically depicted as a cladogram, is explained by a phylogenetic, i.e. transformational process (Nelson and Platnick, 1984). The hierarchy of types represents a logical relation, not a hierarchy of organisms: the fish is not only a vertebrate - it is also a fish. "The vertebrate" as such does not exist in the sense of a reproductive entity or community, reproduction being a *sine qua non* for transformation. The Vertebrata is an abstract concept denoting the logical relation of homology as obtained by downward classification. The same holds for the subtypes, for less inclusive groups contained within the Vertebrata. The Vertebrata, the Tetrapoda, the Sauropsida are notions denoting abstract relations of form which imply a type of organization or *bauplan*. In the context of an evolutionary explanation of the branching order of life, the type or *bauplan* may serve as a guide to hypothetical ancestry: groups pairing off at a dichotomy in the cladogram may be logically derived from a hypothetical ancestor which will have to represent the appropriate type of organization. But the type or *bauplan* cannot in itself represent this ancestor. Lovtrup (1977: 52) missed this point when he states: "evolution has progressed from the apex to the base of the phylogenetic hierarchy". The statement equates the phylogenetic process with the logical analysis of relations of similarity by downward classification, with the effect that classes must have originated prior to the included orders, orders must have given rise to families which in turn originated before the included genera etc. As Schoch (1986: 162) put it. "The ancestral taxon cannot logically include the descendant taxon within it...the Vertebrata did not give rise to the Mammalia, although some organism or species that was a vertebrate gave rise to the Mammalia and all mammals are vertebrates".

If von Baerian types cannot be ancestors, the question may be asked whether Haeckelian groups, arranged in a linear hierarchy, might not fulfill that role? The discussion of the problem of objectification of an ancestor must await the exposition of the methodology of objectification of the genealogical hierarchy (chapter 6). Nevertheless, the work of E. Serres and of E. Geoffroy Saint-Hilaire shows that recapitulation by terminal addition does indeed appear to lend itself to an evolutionary explanation: the serial hierarchy appears to link the ancestor with its immediate descendant.

Continuity versus Discontinuity

It has been argued above that Haeckel, in his "crystallography of the organic" adopted an atomistic background which was in accordance with his materialistic view of life. The same is true for Darwin and his theory of pangenesis. The atomistic philosophy can be traced into modern Darwinism (Rieppel, 1986a), which treats genes and characters as material "atoms" combining to form the organism. The unity of type is treated in a material sense rather than as abstract relation of form, its constituent elements representing substances each with a positional and genetic identity, and their homology is thought to indicate material bonds of inheritance rather than "hidden bonds" as Darwin used to characterize the essence of "idealistic morphology" (Webster, 1984). "On my theory, unity of type is explained by unity of descent", claimed Darwin (1859 [1959: 176])

Since by this conception of organization variation is fundamental, "everything is possible" (Webster, 1984: 208): "...if it could be demonstrated that any complex organ existed which could not possibly have been formed by numerous, successive, slight modifications, my theory would absolutely break down" (Darwin, 1859 [1959: 162]). Darwinian atomism is related to the notion of gradual in the sense of continuous transformation: "...numberless intermediate varieties linking most closely all the species of the same group together, must assuredly have existed..." (Darwin, 1859 [1959: 153]). Continuous transformation by terminal addition to ancestral ontogenies must result in a serial arrangement of ancestors and descendants bridging all intermediate stages between any two levels of organization. This 'way of seeing' emphasizes continuity, but at the same time the clearcut demarcation of levels of organization within the serial hierarchy of life becomes impossible. Darwinian nominalism, discussed in relation to the paradox of the evolving species (chapter 2), is the logical consequence. The linear classification of organisms lends itself to explanation by the evolutionary process, as it links the ancestral segment of the unbroken genealogical nexus directly with the descendant segment, the two gradually, i.e. continuously merging into each other. If, in a series of forms, all but the terminal taxa are privative, it follows that the relation of ancestry and descent is one of simple addition of new characteristics. The serial arrangement of organisms is committed to the principle of continuity and emphasizes process thinking (Rieppel, 1985b).

This is quite a contrast to the hierarchical 'way of seeing' advocated by von Baer. In this view, the deviation of ontogenetic pathways results in a hierarchy of groups within groups which are separated from each other by morphological gaps. The hierarchical 'way of seeing' emphasizes the pattern of order in nature, i.e. the discontinuity between groups which is believed to reflect constraints imposed by a common principle of form within groups. Thus, while the various groups are clearly demarcated from each other at the adult stage and stand in a subordinated hierarchical relation to each other, there follows by implication that there are no transitions between them at the fully differentiated, i.e. adult stage. If such a hierarchy is to be explained by the process of evolution, this can only be done with reference to the epigenetic concept of deviation, which creates morphological gaps by modifications introduced into early ontogenetic stages. The degree of similarity (explained by relatedness) between groups is determined by the succession of deviation events: more closely related groups will share a common developmental pathway which is longer than that shared by less closely related groups. Groups within groups may be reduced to a common principle of form determined by the shared developmental pathway. The hierarchy of types, reflecting the succession of deviations within a given type, corresponds to a hierarchy of common ground-plans of organization from which related groups may logically be derived. The type is nominally characterized by the topological relations of its constituent elements, but materially determined by the developmental program determining its differentiation. From this point of view, homology may have to be based on the developmental parameters producing the type rather than on the type concept itself which represents a logical abstraction of relations of form (Roth, 1984). There follows from a continuity of underlying causes, determining the continuous developmental process, a discontinuity of effects, i.e. disparate morphologies.

Atomism as opposed to epigenesis are alternative 'ways of seeing', one emphasizing phenotypical continuity, the other discontinuity; this results in a serial as opposed to a subordinated hierarchy of life. The crucial point, however, is that all continuity reveals itself to human perception as a more or less finely graded series of steps. An infinite number of points is potentially contained in any line, and consequently an infinite number of segments is potentially contained within any series of forms. Conceptualization of a continuous series of forms will necessarily result in its subdivision in order to establish segments which can be compared and thereby related to each other. In comparative biology, a transformation series is termed a morphocline (Maslin, 1952), the conceptual analysis of which requires its segmentation into a series of character states, linked to each other by a hypothesis of transformation. As each continuous series of forms can only be conceptualized in terms of a chain of discrete segments, it follows that every linear hierarchy can be translated into (or: is complementary to) a subordinated hierarchy containing as many dichotomies as there are links in the chain.

An example for this relation is provided by the early classification of snakes (Rieppel, 1987b). As mentioned before (chapter 3), Cuvier, Brongniart and Duméril agreed on a dichotomous classification of reptiles, separating lizards from snakes. However, it had long been known that some characters such as the reduction of the limbs, but also the jaw apparatus, would lend themselves to establish a gradual, i.e. continuous transition between lizards and snakes (the latter including the amphisbaenians at that time). In fact, Duméril (1803: 346) created the name *Ophisaurus* out of this dilemma: "The name I use to designate this new genus evidently indicates that it must include reptiles which are intermediate between the snakes and the saurians...". Oppel (1811), it had been noted, drew the necessary consequences and united lizards and snakes within a single group, the Squamata. He justified his move by reference to the fact that lizards, amphisbaenians and snakes form so continuous a series of forms, in particular with respect to the reduction of the limbs, that any demarcation of lizards from snakes would have to be highly subjective. Although his classification was accepted by most later authors (the Squamata are today considered to represent a highly corroborated monophyletic group), his new classification of reptiles only provoked the problem of the proper (i.e. "natural") subdivision of this new group: as if the hydra had grown another head! Many solutions were proposed during the first half of the 19th century, but viewed from a modern perspective it was Merrem (1820) who first correctly separated limbless lizards from snakes (which were still considered to include the amphisbaenians, however) in the continuous series of squamatan reptiles. It is worth recalling his uncertainty as to the validity of his classification, however, since this not only does justice to the historical facts, but beyond that illustrates the dilemma of a serial versus a dichotomous arrangement. Not knowing for sure how the series should properly be subdivided, Merrem (1820: xii) asked "whether I was wrong as I subdivided the Squamata into three tribes...or whether I should have left the three tribes together, or whether I should have assumed the existence of five tribes...because...there are transitions which appear to render all subdivision impossible". All this boiled down to 'a way of seeing'. Schlegel (1837) opposed the dichotomous classification of Brongniart (1800), separating lizards from snakes, to the serial arrangement of Oppel (1811), including both within the Squamata, and found both hypotheses of grouping "equally susceptible to defense and opposition": it is a question of *"manière de voire"* (Schlegel, 1837, vol. I: 2).

A continuous series of forms, perceived as a series of discrete steps, can be translated into the axis of a subordinated hierarchy containing as many dichotomies as there are transitions between character-states, and vice versa. A temporalized reading of the *scala naturae*, representing the gradual phylogenetic ascent to ever-higher levels of organization, is complementary to the axis of a branching tree of life which bears as many nodes as there are discrete steps in the ladder of life. The complementary relation of serial versus subordinated classification, of exclusive versus inclusive taxa, shows that the emphasis of continuity versus discontinuity, of atomism versus epigenesis, are likewise complementary 'ways of seeing', which reflects the fundamental complementarity relating pattern to process.

CHAPTER 5

THE FOCAL LEVEL OF ANALYSIS

Serial and subordinated classifications are complementary and hence convertible into one another. This is the reason why, in the present context, the notion of 'hierarchy' is applied to both ways of looking at nature. The serial classification represents an exclusive hierarchy which can be converted into a subordinated hierarchy by the specification of inclusive taxa. The serial hierarchy is compatible with an explanation by a gradual, i.e. continuous process of transformation, whereas the subordinated hierarchy emphasizes discontinuity, expressed by gaps in the morphological continuum. Patterson (1984) has made the point that the term 'hierarchy' should be restricted to the dichotomous subordinated structure of a branching diagram. Such a restriction of the notion of hierarchy introduces an operational criterion into the definition of the term in the sense that it must correspond to that methodology of classification which specifies the mutual relations of groups under groups. A similar understanding of the term seems implied in two recent books (Eldredge, 1985; Salthe, 1985). This difference in the meaning and application of the term 'hierarchy' should be born in mind in the ensuing discussion.

Along a different line of reasoning it has been argued that pattern analysis must be independent from assumptions about process, that is: the analysis of order in nature must precede its explanation by the hypothesis of evolution (Rosen, Forey, Gardiner and Patterson, 1981; Brady, 1985). Two ways of pattern reconstruction have traditionally been advocated in comparative biology, the transformational versus the taxic approach (Eldredge, 1979; Patterson, 1982). The transformational approach seeks to document change of structure by the construction of morphoclines (Maslin, 1952), a representation of graded similarity of form which, if its polarity can be determined, is held to depict the evolutionary transformation of the characters under analysis. In more recent times, the transformational approach, dubbed idealistic by Patterson (1982; see also Dullemeijer, 1974), has come under attack because of its dependence on *a priori* hypotheses of transformation which themselves are unobservable; hence, transformational pattern analysis cannot be independent from hypotheses about process. The validity of any (explanatory) hypothesis about process can only be evaluated (tested: Brady, 1985), however, if it is contrasted with a pattern (to be explained) which has been reconstructed independently. If pattern analysis were based on or influenced by assumptions about the process of evolution, order in nature would no longer constitute independent evidence against which evolutionary theory could be evaluated: any such test of evolutionary predictions against pattern analysis would become circular. It follows that the transformational approach to pattern analysis is untestable. The same conclusion follows from the fact that transformational homologies are to a large degree untestable.

This is the reason why the taxic approach has gained preference over the transformational approach. The taxic approach focuses on patterns of organismic diversity over space and time. It reconstructs the subordinated relations of groups within groups on the basis of logical analysis of character distribution. The taxic approach represents a conceptual framework which permits the representation of homologies in a subordinated pattern. It, too, does "not neutrally 'represent' data [homologies] but adds the hypothesis of [subordinated or inclusive] hierarchical order" (Brady, 1983: 50). Homologies specify inclusive taxa the members of which may be interpreted as being related by common descent from a hypothetical common ancestor. The inclusive taxa of the subordinated hierarchy represent logical relations of form or similarity - they do not indicate ancestral groups. By its evolutionary interpretation, the branching diagram is transformed into a phylogenetic tree, its logical relations are explained on the hypothesis of descent with modification. The designation of actual ancestors, by forcing down a terminal taxon to a branching point (node) in the cladogram, converts segments of the latter into a serial hierarchy.

It is the object of the present chapter to discuss the ontology and objectification of the genealogical hierarchy, two aspects which require the specification of the focal level of analysis.

The Hierarchy

The serial or exclusive hierarchy, linked as it is to an atomistic background, emphasizes transitions between levels of organization in an upward direction. The atomistic background implies a reductionist perspective: the understanding of life starts with its building blocks which combine to produce phenomena of ever higher levels of complexity. Or, to put it the other way around: the explanation of the phenomena of life requires their atomization, i.e. the reduction of organisms to their constituent elements, organs or organ complexes. These in turn are composed of cells, which contain the genetic information for their appropriate differentiation and collaboration in tissues of various functions. The organism, its appearance and relations, are determined by upward causation. Genes control the synthesis of enzymes and proteins, which determine the functions of tissues, which combine to form organs, which in turn group together to form the organism. Darwin explained the generation of variable offspring, the raw material of evolution, on his theory of pangenesis, reducing embryogenesis to the fortuitous aggregation of particles or *gemmulae*. "Evolution from Molecules to Men" is the slogan of Neo-Darwinism (Bendall, 1983).

The subordinated or inclusive hierarchy, in contrast, emphasizes distinctiveness and discontinuity between levels of organization. It represents, as Eldredge (1985: 139) has put it, "an alternative way of looking at nature", stressing the individuality and emergent properties of the various levels of the hierarchy of life. It is not claimed that this is necessarily the only one and "correct" way of looking at nature, but only that the hierarchical or taxic approach might prove more useful heuristically, i.e. be of greater explanatory value (Eldredge, 1979; 1985: 11). Phenomena at any level may not only be caused from below, but also from above. Properties characterizing a given level of organization do not simply result from the sum of its parts; they are not explained by reduction to lower levels of organization only. Instead, causation from below combines with causation from above to produce effects which determine emergent and unique properties imparting individuality on the organized and coherent entities at any hierarchical level (Eldredge, 1985; Salthe, 1985). Eldredge and Salthe (1984) have called 'constraints' those properties of a lower level which affect processes at the next higher level, while a higher level influencing the next lower level by downward causation provides the 'boundary condition' for events at this lower level.

The individuality of each hierarchical level and its emergent properties resulting from the unpredictable interaction of lower level initiating conditions and constraints with upper level boundary conditions (Salthe, 1985:101) makes it necessary that the analysis of processes be related to a specified focal level of analysis. As processes depend on the individual properties of a given hierarchical level, they cannot be explained by straight reduction to lower levels of causality, so it is claimed. A simple and much debated example, to be more fully discussed below, is the concept of species selection (Stanley, 1979). Assuming that the specific diversity in the biota is conditioned and constrained by natural selection, the question may be asked how selection operates to check on species diversity. A reductionist explanation would consider species diversity to be the effect of intrapopulational (i.e. Darwinian) selection, acting between individuals and thereby causing phenomena at a higher level of the hierarchy. If it is admitted, however, that species are individuals with emergent and unique properties, the problem of species diversity might become one of species selection: species act as individual entities in a selective process relating to their emergent properties and taking place at their proper hierarchical level (Eldredge, 1985). To talk about species selection presupposes knowledge of what the species concerned are, or, in other words: to talk about focal level processes presupposes knowledge about the structure of various biological hierarchies. The hierarchical 'way of seeing' therefore requires a discussion of the ontology of natural entities as well as of the structure of biological hierarchies. To start with the latter, Vrba and Eldredge (1984) and Eldredge (1985) distinguish five inclusive hierarchies relevant to the evolutionary process, to which the hierarchy of ontogenetic differentiation will be added. Some of these hierarchies were already implicit in the material presented above; the present paragraph was opened with a delineation of the somatic hierarchy (Vrba and Eldredge, 1984: 150): it runs from proteins, organelles and cells through tissues, organs and organ systems up to the phenotype of the individual organism.

The somatic hierarchy is to some degree paralleled by the hierarchy of the epigenetic process of ontogenetic differentiation as expounded by von Baer (1828; see chapter 4): his "primary" and "histological" differentiation produce the cells and tissue precursors of the "morphological differentiation" producing the organs and organ systems shaping the adult body.

The epigenetic process of ontogenetic differentiation is paralleled by yet another hierarchy, namely "that of homology, the hierarchy of evolutionary novelties" (Vrba and Eldredge, 1984: 150). As has been discussed above, epigenetic deviation results in a developmental process progressing from the differentiation of more widely distributed features characterizing more inclusive groups to the development of less widely distributed characters diagnosing less inclusive groups. The characters that diagnose groups are homologies (Patterson, 1982), and if the relation of homology is explained by common descent from a hypothetical ancestor, the hierarchy of homologies determines the hierarchy of evolutionary innovation.

The hierarchy of homologies in turn is congruent with the taxic hierarchy (Eldredge, 1985: 140) which may be expressed in Linnean terms (Vrba and Eldredge, 1984: 150) if the progressively less inclusive taxa of the Linnean hierarchy are diagnosed by homologies at the correct level of their inclusiveness: the pentadactyle limb is characteristic of mammals, yet it diagnoses a more inclusive group, the Tetrapoda (Patterson, 1982).

To these four hierarchies, two additional and perhaps the most important ones recognized by Vrba and Eldredge (1984; Eldredge, 1985) must be added, viz. the genealogical and the ecological hierarchies. The genealogical hierarchy runs from codons and genes through organisms and demes to species and monophyletic taxa (of higher rank) (Eldredge and Salthe, 1984). Eldredge and Salthe (1984) have expanded the ecological hierarchy so as to overlap in its less inclusive levels with the somatic hierarchy as discussed above: it then runs from enzymes and cells through organisms and populations to local ecosystems and biotic regions. This procedure permits the parallel arrangement of the genealogical and ecological hierarchies and therewith the analysis of interactions between the two. The schema given by Eldredge (1985: 165, table 6.2.; reproduced below) shows the individual organism to partake in both hierarchies at the same relative level. They play different roles in different contexts, but their presence in both hierarchies indicates a minimal interaction between the two at least at the organismic level (Eldredge, 1985: 188).

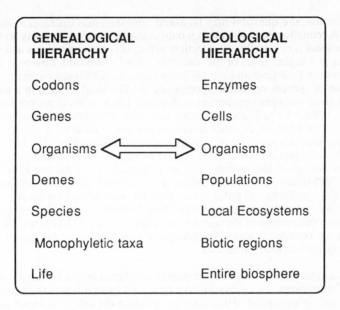

Fig. 9: The genealogical and ecological hierarchies interact minimally at the level of the individual organism (after Eldredge, 1985).

A most important aspect of the genealogical versus ecological hierarchies as drawn up by Eldredge (1985) is the point that the species does not appear as a distinct level of organization in the ecological hierarchy. This could be a key to the potential solution of the paradox of the evolving species (as treated in chapter 2), which originates from the extension of the biological species concept over time and space. It has been noted above that in spite of the title of his book, Darwin did not really address the problem of *The Origin of Species*. This is because he was primarily concerned with the causation of evolutionary processes. If the cause is a continuous one, causing gradual, i.e. continuous, evolutionary change, all possible intermediates obtain between any two structures. Species then can no longer be objectified through time and space. If, however, the reconstruction of the pattern of species diversity over time and space treats species as discrete entities in the hierarchy of nature, diagnosed by species-specific traits, they can no longer evolve - or evolution must be discontinuous, i.e. saltatory at the phenotypic level.

The ecologist focuses on the mechanisms of the evolutionary process, and hence moves up and down in the ecological hierarchy. Research concentrates on populations rather than on species, i.e. on actually interbreeding groups of organisms which occupy a certain space at a given time. The species as a natural entity, preserving its coherence and individuality through space and geological time drops out of focus. The phylogeneticist, on the other hand, is concerned with the reconstruction of the pattern of order in nature to be explained on the hypothesis of the evolutionary process. Phylogenetic research will be related to the genealogical hierarchy. Species are recognized as a distinct level of the genealogical hierarchy, objectified through space and time by the shared presence of species-specific characteristics.

The basic question to be asked is whether the species does indeed represent the basal unit of evolution, as would be implied in the title of Darwin's book. Studies of population genetics have made clear that mechanisms of Darwinian evolution operate at the level of actually interbreeding and hence localized populations, rather than at the level of species. In a modern textbook on the subject, Ayala and Kiger (1980) characterize evolution "at the genetic level" as consisting "of changes in the genetic constitution of populations": "In evolution, the relevant unit is not an individual [organism] but a population. A *population* is a community of individuals linked by bonds of mating and parenthood" (Ayala and Kiger, 1980: 598).

The notion of "population", however, can designate a group of interbreeding organisms at various levels of inclusiveness. "A *local population* is a group of individuals of the same species living together in the same territory" (Ayala and Kiger, 1980: 598), but: "boundaries between local populations are often fuzzy" (Ayala and Kiger, 1980: 598). Therefore: "The most inclusive Mendelian population is the species…" (Ayala and Kiger, 1980: 598). From the fuzziness of boundary conditions determining the inclusiveness of the population there results some ambiguity as to which level should be considered the most relevant one in the study of evolutionary mechanisms. Ayala and Kiger (1980: 598) designate the "population" as the "relevant unit" in evolution, while species are considered "independent evolutionary units…".

The focal level of analysis of the evolutionary process must therefore be determined by the invoked causal mechanisms rather than by theoretical considerations of what a population is or should be. The mechanisms of evolutionary change invoked by Ayala and Kiger (1980: 628) are natural selection and genetic drift.

Natural selection (in fact: Darwinian selection; see below) is defined by Ayala and Kiger (1980: 658) as "…the differential reproduction of alternative genetic variants…". This definition presupposes actual interbreeding within a population localized in time and space. "The ultimate outcome of natural selection may be either the elimination of one or another allele…or a stable polymorphism with two or more alleles" (Ayala and Kiger, 1980: 661). That this process operates at the level of localized populations is borne out not only by the differentiation of geographic races ("genetic distinct populations of the same species"; Ayala and Kiger, 1980: 719), but also by so-called "area-effects" ("…areas in which the genetic composition of the population is strikingly different from that in the surrounding contiguous area"; White, 1978: 154). Needless to say that the test of natural selection by experiment is likewise confined to closed (i.e. localized) populations.

Genetic drift, the second mechanism of evolutionary change invoked by Ayala and Kiger (1980: 643), refers to changes in allele frequency in a "certain population", and its effect is larger the "smaller the number of breeding individuals in a population" is. It therefore depends on "the effective population size" (Ayala and Kiger, 1980: 645), and hence on the number of actually interbreeding individuals.

In summary, genetic and ecological mechanisms of change operate at the level of localized populations; if migration occurs between such populations, its evolutionary impact is again on localized entities. To put it in other words: whatever the role of the species in the genealogical hierarchy is, the origin of species diversity must be explained by genetic processes operating at the level of localized populations. The appropriate research program is related to the ecological hierarchy.

Expanded through space and geological time, the problem of the origin of species diversity can be addressed from two different viewpoints. The transformational approach seeks to document, within an evolutionary lineage, the continuity of morphological change. "Morphological continuity on a broad scale is the principle evidence favoring evolution as a historical explanation for the diversity of life" (Gingerich, 1985: 28). The level of change is again the population: "Closely spaced samples of former *populations* preserved in the fossil record are critical for testing the continuity or discontinuity of morphological evolution within and between species" (Gingerich, 1985: 30; emphasis added). The transformation of species is thus documented by change taking place at the level of populations which continuously merge into each other within the unbroken, i.e. continuous genealogical nexus (Mayr, 1963, 1971). The transformational approach lies at the heart of the concept of phyletic transformations of "species", i.e. the (gradual, continuous) transformation of "species" as a whole through time. The term species is put into quotation marks because "species" succeeding each other cannot be objectified except by a largely arbitrary subdivision of the string of successive generation on the basis of the degree of morphological change.

Mechanisms of population genetics discussed above would predict, however, that the wholesale transformation of species is rare to occur (Dobzhansky, 1951; Mayr, 1963), although it is possible given enough time and strong enough directional selection pressure. The so-called "founder principle" (Ayala and Kiger, 1980: 651) relates to the fact that genetic change, and consequent morphological change, is more likely to occur in relatively small and localized populations or demes, i.e. in relatively isolated subunits of species. These effects of the mechanisms of population genetics do not prevent the fairly rapid phyletic transformation of entire species, however, if these are forced through some temporary bottleneck which reduces the species to a small and localized population. The bottleneck may be environmentally induced. Willmann (1979) has shown that rapid morphological change in island species of Miocene freshwater snails occurs synchronously during periods of marine transgression, when the freshwater habitat is drastically reduced. A similar model has been invoked in the discussion of morphological change in molluscs from Cenozoic deposits of the Turkana basin in northern Kenya (Williamson, 1981, and sequels 1982, 1983). The flush and crash model of speciation proposed by Carson (discussed in White, 1978: 110-111) may result in a special case of a bottleneck-situation, induced by intrapopulational mechanisms. Stanley (1979: 173) has made the important remark, however, that although bottleneck situations may well account for phyletic transformation of entire species involving significant morphological change during short periods of time, this evolutionary mechanism does not increase the *diversity* of the number of species in time and space! Moreover, the bottleneck model of speciation presupposes the reduction of the whole species to a localized population and hence, once again, is related to the ecological hierarchy.

The pattern of diversity of species through time and space is the object of the taxic approach. This approach is linked to the allopatric or parapatric speciation model (and its special cases such as the founder principle) as advocated by Mayr (1942, 1963) and incorporated into the punctuated equilibria model of evolution (Eldredge and Gould, 1972; Gould and Eldredge, 1977). The term 'speciation' is restricted to events which produce new species from existing ones, thereby implying some branching event and hence adding to the taxic diversity of the biota (Stanley, 1979: 173). The model rests on the assumption that evolutionary change will be confined to a small and relatively isolated population of a given species which, as a consequence of changes in its local gene-pool, will branch off from its mother species. Homeostatic mechanisms, ultimately of genetic origin and usually visualized to determine developmental constraints, are invoked to explain the observed stasis within species throughout their duration, as documented by the fossil record. The homeostatic mechanisms are believed to break up by means of a "genetic revolution" in small, peripheral isolates, with the effect that stored genetic variability may become expressed at the phenotypic level, and new variability created by various sorts of mutations. An experimental phase sets in at the genetic and, as a consequence thereof, at the phenotypic level, eventually causing the reproductive isolation of the daughter species from its mother species. The expression of stored genetic variability or of new mutations, as well as genetic drift, may induce the creation of a new form. If the new morphology proves viable in the given environment, if it is not eliminated by (interspecific) species selection, it may be perfected by (intraspecific) Darwinian selection. The theory of punctuated equilibria thus represents an expansion, not a replacement of Darwinian evolution. Starting out from the distinctiveness and individuality of species, it treats speciation in peripheral isolates as introduction of variation at the species level, and species selection as differential sorting at this focal level. The taxic approach again invokes mechanisms of change operating at the level of localized populations in order to explain the *origin* of species diversity. In addition, however, it expands the theory of evolution to biotic (density-dependent) interactions between species, and by the concept of species selection it invokes differential sorting at the species level. By the taxic approach, the study of evolutionary process is thus expanded from the ecological to the genealogical hierarchy. This is a claim which requires further scrutiny in the following paragraph, since it is based on a new ontology, treating species as individuals.

It should be noted that the theory of punctuated evolution must, like any other theory of evolution, start out with the reconstruction of pattern (which is to be explained by that theory). The model of allopatric speciation is invoked by the proponents of the theory of punctuated equilibria to explain the observed stasis within species and the morphological gaps between species. The theory rests on the diagnosis of species through time and space by means of species-specific characters (*differentiae:* Eldredge, 1979), with the effect that the so-called stasis within species is often restricted to these characters only (Charlesworth, Lande and Slatkin, 1982). Conversely, these *differentiae* cannot be allowed to change gradually, since if they did, they would lose their function as species-specific traits.

The taxic approach focuses on discontinuities and hierarchical subordination in organic diversity, which is explained by a continuity of underlying causes (Gould, 1985: 2), viz. changes of the genetic constitution of small and localized populations. To the exception of the theory of species selection (see below), it would thus appear possible to restrict the study of the *process* of evolutionary transformation, i.e. the populational mechanisms of genetic and ensuing morphological change, to the ecological hierarchy. Process orientated research is related to a hierarchy of entities which does not include species, and hence it does not have to deal with the problem of species demarcation through time and space. As Eldredge (1985: 158) put it: "Unless a species is restricted to a single population, it is impossible to go to nature, sample a species, and see it as a dynamic interactor in the ecological realm". This, however, is exactly what is required in a study of causal mechanisms of evolutionary change. The taxic or pattern-orientated approach, on the other hand, is concerned with the structure of the genealogical hierarchy, incorporating species as discrete and hence discontinuous entities of nature. That the study of the causal mechanisms of the evolutionary process must be based on observable populations and hence is related to the ecological rather than to the genealogical hierarchy creates a basic conflict between the analysis of process and the reconstruction of pattern. (The same problematical relation obtains with respect to the notion of actual specific ancestry, which turns out to be incompatible with the logic of pattern reconstruction, a relation to be discussed in detail in chapter 6.)

A basic distinction has to be drawn in evolutionary theory. The investigation of actual causes of evolutionary change must be restricted to localized populations amenable to observation and, in certain cases, even to experimentation. The analysis of evolutionary process in that sense is related to the ecological hierarchy. This is a different research program from that which extrapolates, on the principle of uniformity of natural laws, the action of those causes from the present into the past in order to explain the pattern of the genealogical hierarchy on the hypothesis of a phylogenetic process. Evolutionary biology deals with the mechanisms of change as they are observed to operate in localized populations; the hypothesis of a phylogenetic process, of descent with modification, assumes the action of these same forces in the phylogenetic past and thus is based on the extension of causes from the ecological hierarchy to the genealogical hierarchy, from biological time to the phylogenetic time scale.

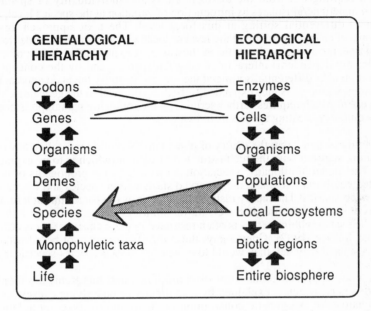

Fig. 10: A mildly reductionist view of some interactions between the genealogical and ecological hierarchies: the analysis of causes of evolutionary change is restricted to the ecological hierarchy; the action of these causes is extrapolated into the genealogical hierarchy in order to explain the pattern of order in nature.

Phylogeny corresponds to a causal explanation for the genealogical hierarchy. In the light of this argument, the species problem can be summed up following Mishler and Brandon's (1987) distinction of criteria of grouping from ranking of groups at the species level. The criteria of *grouping* are relations of homology or synapomorphy: they specify a hierarchy of "groups under groups", as Darwin put it, i.e. the genealogical hierarchy. Some level of that hierarchy may be *ranked* at the species level, and this ranking may be justified, i.e. explained with reference to ecological or developmental causes. If the criteria for ranking are to provide a correspondence to the biological species concept, they must refer to the mechanisms tying an actually interbreeding group of organisms together. The species then becomes coextensive with a localized population (Rieppel, 1986b). This conclusion obviously bears on the concept of species selection!

Species selection, postulated on the basis of a hierarchical approach to evolution (Eldredge, 1985), is the most important hypothesis which transcends this reductionistic view of pattern versus process. Species selection relates to emergent properties imparting individuality on particular species taxa. Particular species taxa appear as discrete entities in the genealogical hierarchy which have a beginning as well as an end in time and which have a unique evolutionary role. As already pointed out (chapter 2), the birth of the species taxon corresponds to its origin from a branching event, whereas its death results either from terminal extinction, or from its splitting into two or more daughter species. The logical corollary is that a mother species splitting off a daughter species becomes itself the sister of its daughter, i.e. a new species, even if it does not undergo morphological change (Ax, 1984; Willmann, 1985). A second corollary is that phyletic transformation is defined away as mode of speciation, which appears to be justified since it does not add to taxic diversity (Stanley 1979). The unique evolutionary role of the species would result from reproductive coherence within it, but reproductive isolation from other species. Taking the above considerations together one obtains the criteria of individuation for species taxa.

The Ontology of Natural Groups

Species as individuals can do something: they can be born, evolve, go extinct. Classes cannot do anything: defined on immutable essences, they cannot change, evolve, come into existence and disappear. "National state" (a class) cannot wage war, "Canada" (an individual), can (Ghiselin, 1987).

The thesis that species represent individuals was reviewed in a previous contribution (Rieppel, 1986b), discussing the paradigms evoked in support of this "new ontology" (Eldredge, 1985), viz. the individual organism, economic theory and *Hydra*. If anything, these paradigms make clear that the notion of individuality entails the preservation of identity in the course of change. The individual organism remains the same, i.e. it preserves its identity in spite of changes of its material appearance throughout the ontogenetic process leading from birth to death. By analogy, species meeting the criteria given above would seem to meet this requirement, too. Originating from a branching event and terminating by splitting (or extinction), the species remains the same and hence identical throughout its duration from one to the next cladogenetic event regardless of any phyletic change taking place in between. If species are individuals, they cannot be defined, but only discovered, described and named (Ghiselin, 1974): species names are proper names, imparting identity to the species throughout the phases of possible phyletic transformation. The thesis that species should be considered as individuals claims independence from any operational criteria of species recognition or discovery. Species are asserted to be real and to exist in nature irrespective of the observer's willingness or ability to recognize them - they exist as units of evolution and selection or, in more general terms, as products of speciation (Ghiselin, 1966, 1981; Hull, 1976, 1978, 1980). The previous review (Rieppel, 1986b) stressed the problems of the objectification of species either as closed, or as open systems and the reader must be referred to this earlier study for details. It was concluded that if the notion of individuality implies identity in the course of change, and if species are to be individuals, these must be objectified as spatiotemporally restricted entities on the basis of some *individual* essence. Viewed from a reductionist perspective, the individuality of an organism throughout its ontogeny from its birth to its death is preserved by its genome. In a similar sense, the individuality of a spatiotemporally restricted species must be preserved by some essential property, usually conceptualized as its genetic makeup determining developmental constraints (the "real" essence according to Webster, 1984) and by them the stasis of species-specific traits (the "nominal essence" *sensu* Webster, 1984, representing the properties on which the species name is based).

It might be argued that if species are individuals philosophically, they cannot have *essences* in the Aristotelian sense, which were after all properties of Aristotelian classes. This argument can be countered with reference to the essentialistic species concept expounded by Aristotle. He believed species to be immutable, representing elements of the 'world of being'. The immutable form of the species would continuously become actualized during the reproductive cycle, i.e. by the continuous (ontogenetic) 'becoming' of individual organisms. The species-specific form was conserved during the reproductive cycle by the soul which, as the "principle of knowledge", would "guide" the ontogenetic development of individual organisms. The soul made it possible for the individual organisms, elements of the 'world of becoming', to partake in the eternity of the species. Or in other words: every member of the species shared in the same essence, i.e. it shared the same "principle of knowledge" of form. The same can be said of a genetically determined developmental program, maintaining homeostasis and hence historical as well as spatial cohesion within species, imparting on it identity through time and space.

In cladistics not only species, but even more inclusive taxa are considered as individuals, provided they have been objectified as monophyletic groups (Patterson, 1982; Eldredge and Novacek, 1985). The objectification of a monophyletic taxon *qua* individual must depend on the demonstration of unique features (autapomorphies) diagnosing the group in question. These, however, must have been created by some deviation of the ontogenetic process, effecting what von Baer (1828) called the "individuation" of a taxon. Autapomorphies therefore represent the "essential" characteristics objectifying the individual taxon: they represent the "essence" of individualized taxa.

It might further be argued that the notion of an 'individual essence' confounds logical and historical relations. Species as well as higher taxa are categorized as individuals in the sense which this notion bears in the context of logics (Ghiselin, 1974: 536). As a consequence, there is no need nor any possibility for any such thing as an 'individual essence', thought to impart identity on the species through space, time and phyletic change of appearance. In reply to this argument it must be asked whether 'species' and 'monophyletic taxa' are to represent a logical concept (as does the type concept expounded by von Baer, 1828), or whether they are thought to represent groups of organisms interacting in the ecological and phylogenetic realm, i.e. in spatial and historical dimension?

In the present context another point will be emphasized, namely how it should be possible to invoke speciation as an argument supporting the reality of species taxa before a hypothesis of relationship between two particular species has been formulated: "This looks like putting the cart before the horse" (Rieppel, 1986b: 298).

The thesis that species (Ghiselin, 1974), and by implication monophyletic taxa of supraspecific rank (Eldredge and Novacek, 1985), are individuals has since it original proposal become embedded in the taxic approach to pattern analysis (Patterson, 1982). Species (and monophyletic groups of higher rank) are not defined, as are classes; rather, they are discovered (recognized), named (with proper names) and related. The main criteria of individuation of species are, as has already been noted, spatiotemporal restrictedness (species have a beginning and an end in time, i.e. they are born and die) and organization among the parts (by means of reproductive coherence). It is clear that if species really *are* individuals, this ontology would "force us", as Eldredge (1985: 126) put it, to adopt the taxic approach: "reorganization of our ontology of such biological 'things' seems automatically to imply hierarchy" (Eldredge, 1985: 126). Treatment of species (and supraspecific taxa) as individuals imposes upon us a taxic 'way of seeing'.

However, it is not quite clear how it should be possible to make absolute statements on the ontology of species which assert their reality beyond the intentions and constraints of human cognition. It is unclear how the theory of evolution and models of cladogenetic speciation should support the call for a reality of species taxa when the objectification of species as units of evolution meets with serious epistemological difficulties. "...then so much worse for the epistemological perspective", is Hull's (1987: 176) laconic answer: "Species may not *seem* like genealogical units, but if they are to evolve, this is precisely how they must be construed" (Hull, 1987: 176). The "new ontology" thus forces a view on us (Eldredge, 1985) which is derived from a theory (of evolution by natural selection) that should in fact follow the observation of species rather than precede it (Brady, 1982, 1983, 1985). It has been argued, however, that since all knowledge is hypothetical, this view "forced on us", i.e. the taxic approach, need not necessarily represent the "correct" and only possible one, but it may represent the heuristically most rewarding one. What, then, are the heuristic advantages of the taxic approach, combined with a revised ontology of natural entities? No doubt its greatest advantage is the conquest of pure reductionism (Eldredge, 1985; Salthe, 1985). Upward causation is complemented by downward causation or, in other words and with reference to Fig. 10 above, arrows can be assumed to be effective in the opposite directions also. In fact, Eldredge (1985: 187) has sketched the "hierarchy purist's dream" by conceptualizing reciprocal interactions not only up and down between all the levels of individuality of the genealogical and ecological hierarchies, but also between any two individuals of these two hierarchies. The collapse of a regional ecosystem may not only affect the individuals of the genealogical hierarchy (for instance by extinction), but reciprocally the origin of new species and therewith of new supraspecific taxa may have profound effects on populations embedded in a regional ecosystem or even beyond. The taxic approach, incorporating the new ontology, permits the expansion of process analysis from the ecological to the genealogical hierarchy, and it accounts for reciprocal interactions between the two.

To put it in more simple terms, the taxic approach permits the extrapolation of the theories of variation and selection from the level of the individual organism to the level of the individual species (Gould, 1980; Vrba and Eldredge, 1984; Eldrege, 1985). Looking back at the basic tenet of Darwin's theory of selection, it is readily appreciated that evolution was postulated to be driven by a mechanism of input of variation followed by differential sorting. Darwin restricted causal mechanisms of evolution to the level of the individual organism as part of a reproductive community embedded in its local environment. Variation is introduced into the gene-pool of the population by mutations in the genome of individual organisms. Natural selection results from differential sorting among the organisms constituting the population with respect to some (biotic or abiotic) parameter of fitness.

Taking the species to represent an individual, it is possible to extrapolate the theory of natural selection to the species level, resulting in what has been called "species selection" (Stanley, 1975, 1979; see also above). Species birth, i.e. speciation processes would introduce variation between species which is considered to be random with respect to any long-term trend within species (Vrba and Eldredge, 1984: 164) as well as with respect to environmental parameters. Instead of being guided by Darwinian selection in an adaptive sense, speciation processes thus entail a strong stochastic component (Gould, Raup, Sepkoski, Schopf and Simberloff, 1977). Variation between species represents the raw material for differential sorting or selection among species with respect to environmental conditions, including biotic and abiotic parameters. The theory of macroevolution is thus based on an extrapolation of the Darwinian theory of selection to hierarchical levels above the individual organism: variation is introduced at the species level by stochastic speciation events, while differential sorting affects species as individual (in the original sense of the word: indivisible) units. The corollary is that "species should vary in characters that enter at the species level and that such emergent characters are the cause of species selection" (Vrba and Eldredge, 1984: 164).

Species selection is a process which, if it deserves recognition as a mechanism different from Darwinian selection, must affect emergent properties of species resulting from unpredictable interactions of lower level initiating conditions and constrained by upper level boundary conditions. It is quite obvious that if species selection does represent a higher level mechanism of evolution, the recognition of macroevolutionary events represents a valuable and necessary amendment of the Darwinian theory of evolution (Gould, 1980). On the other hand it is equally obvious that, while compatible with the species *qua* individual thesis, the theory of macroevolution must be couched in the vocabulary of the taxic approach. Kawata (1987:425) has argued that "Species selection by differential extinction and speciation usually plays a role mainly in the description of evolutionary trends, and is not understood to be a causal process". According to this argument, species selection corresponds to a description of evolutionary trends expressed in taxic terms, invoking differential sorting (i.e. extinction) among species the *cause* of which "does not lie at the species level" (Kawata, 1987: 425-426). The point is, that species selection, according to Kawata (1987: 426), is only a "descriptive theory unless species are defined as actually interbreeding populations". In other words: the sum of individual deaths of organisms in an actually interbreeding population is the proximate evolutionary *cause*; its *effect* (Vrba, 1983) may be the extinction of the species. It is therefore vital for the concept of species selection that there really exist emergent species-specific properties which may be subject to selection. Unless this can be conclusively demonstrated, the theory of species selection remains a descriptive concept related to causes acting at the level of localized populations within the ecological hierarchy.

A similar line of reasoning may be read into Sober's (1984) distinction of selection *of* individual organisms (embedded in an actually interbreeding population) *for* traits that may have effects at some higher hierarchical level. Sober has made it quite clear that the demonstration of selection *for* "species-wide properties" (Sober, 1984: 365) is a necessary but not a sufficient condition for the concept of species selection. After all, one can have selection *for* group properties, yet the objects *of* selection remain the individual members of a reproductive community (Sober, 1984: 279). The requirement therefore is that there be selection *of* species *for* species-wide properties.

One, and perhaps the main, problem of the arguments about macroevolution, the hierarchical approach to evolutionary theory, and the implied individuality of species results from the threat of confusion of the *explanandum* and *explanans* (Brady, 1985). Mayr (1987a: 156) concludes from the "fact of their evolution" that species can have no essence. How does he know, if the evolution of species is inferred from a hypothesis of genealogical relationship between similar organisms? On the other hand, Mayr (1987a) is bound to run into problems with the spatiotemporal restrictedness of species (see also chapter 3: the paradox of the evolving species), if he does not allow their parts or members to share in a common essence. This is why he wants to replace the notion of individuality by that of population with respect to species. The "populational thinking" advocated by Mayr (1982), coupled with his rejection of essentialism, is a result of his interest in the causal mechanisms of evolutionary change taking place within and between populations, i.e. within the ecological hierarchy. In a similar sense, Ghiselin (1987: 132) asserts that "...scientific classifications are etiological, or causal, not phenomenal, or based on superficial appearances". This claim must be based on an extrapolation of causal theories of evolution as explanation for the observed hierarchy of homologies! The claim that one species descended from another must be based on the observation that one group of organisms is more similar to some other such group than either of the two is to a third (with respect to any character, be it morphological, biochemical, behavioral, etc.). It therefore seems impossible to agree with Eldredge (1985) and others that it is ontology which forces us to consider species as individuals, only to find this corroborated by a (subordinated) hierarchical pattern of order in nature. It was argued above that the taxic approach, the theory of punctuated equilibria and its incorporation into a theory of macroevolution (Gould, 1985) depends critically on hypotheses about process, i.e. on the assumption of the prevalence of allopatric or parapatric speciation. It is not ontology which forces us to look at nature in a certain way: this argument mistakes "a psychological conjunction for a logical tie" (Kitcher, 1986: 649). Instead it is the regularity (if it prevails) of order in nature which calls for a causal explanation (Brady, 1983), and the only way to find out whether allopatric or parapatric models of speciation have any claim of prevalence over other models of evolutionary change is to contrast these causal explanations with an independently reconstructed pattern.

The models of allopatric or parapartic speciation derive from the study of localized populations. The prevalence of these or other models of speciation in the causal explanation of the phylogenetic past depends on their compatibility with the manner of objectification and reconstruction of the pattern of order in nature. This, at any rate, is the way the whole discussion was started. It was the analysis of the pattern of trilobite distribution through time and space which led to the assertion that species can be recognized at more than one time and in more than one place on the basis of species-specific characters (Eldredge, 1971, 1979) - actually no surprising result in the context of stratigraphic research (Rieppel, 1983b)! Stratigraphy seeks to correlate different strata cropping out at different localities on the basis of the fossils contained within the sediments. Ecological influences aside, a similar fossil content is inferred (Harper, quoted in Schoch, 1986: 211-212) to indicate a corresponding age of the strata under investigation. It is intuitively obvious that such a research program depends on the identifiability of species through time and space. In other words: species which cannot be clearly demarcated from other fossils in the sample are useless for stratigraphical purposes. Levinton (1983: 118) succinctly points out how stratigraphy depends on fossil species "whose recognition is based upon a threshold of perceived phenetic distance". The very fact or convention, however, that species have to be characterized by species-specific traits through time and space not only implies stasis within the species for the duration of its existence, but also discontinuity between species - with respect to these species-specific traits. This was the pattern which was explained by Eldredge (1971) on the assumption of the prevalence of the allopatric speciation model. The rest follows, viz. the theory of punctuated equilibria (Eldredge and Gould, 1972), the emphasis of (subordinative) hierarchy and discontinuity, and the species *qua* individual thesis (Eldredge, 1985): on the basis of pattern analysis, species appear as individualized taxa!

The alternative is to focus continuity and a serial rather than subordinated classification. Species then become arbitrarily delineated (Gingerich, 1979, 1985) segments of the unbroken genealogical nexus. On the principle of continuity of causes determinating continuous and adaptive morphological change (Rieppel, 1987a) , the spatiotemporal restrictedness of species dissolves into a string of evolving populations. It is these localized populations, not the species, which constitute, from this point of view, the basal units of evolution because:

> "Contrary to Ghiselin, individual species are not single populations..." (Stebbis, 1987: 199). "The same problem exists with defining cities in a region such as the northeastern United States. An extreme 'splitter' might recognize every named suburb as a separate city. A 'lumper', looking from an airplane, might decide that the entire mass of buildings...is a single huge city. Intermediate opinions of all sorts could exist. Can anybody say which one is correct?" (Stebbins, 1987: 202).

What is observed are clusters of organisms related to each other by a greater or lesser degree of similarity. It is the method of pattern reconstruction which determines what species can or should be - rather than ontology forcing the view of a certain pattern on us. The taxic approach emphasizes discontinuity and therewith the discretion of hierarchical levels: it is compatible with the thesis that species are individuals, each species being characterized by an *individual* essence, corresponding to the underlying cause of its diagnostic characters, of its *differentia* within the *genus proximum*. The analysis of actual processes of evolutionary change focuses on transitions between groups and hence is incompatible with the notion of spatiotemporal restrictedness of species - prerequisite for species individuality. Again we find ourselves forced back to the incompatibility of the notion of actual ancestry with the logic of pattern reconstruction.

Ontology does not force any particular view on us; instead, pattern analysis may force a certain explanative theory on us which is compatible with certain ontological considerations. Accepting the precedence of pattern analysis over its explanation by process, we are thus relegated to the complementarity of continuity versus discontinuity, of serial versus subordinative classification, of the transformational versus the taxic approach. And instead of submitting to metaphysical considerations on the ontological status of the species, we must ask how pattern - and with it species - is to be objectified, before looking for explanations. If discontinuity and subordination of natural entities is emphasized, species may be considered individuals; species level processes are no longer fully reducible to microevolutionary causation. If, on the other hand, continuity and gradual transitions between what appear to represent natural "entities" are emphasized, species lose their spatiotemporal restrictedness and hence can no longer be considered as individuals; macroevolution appears to result from nothing but a summation of microevolutionary events.

There are different 'ways of seeing', and rather than asserting the superiority of 'one way of seeing' over the other with reference to some reality postulated to exist independent of human cognition, it might prove more interesting to learn about the structure of the world we live in by the investigation of the consequences of objectification of this world from either perspective. "In a trivial, obvious sense all our conceptions are 'intellectual constructs' ", writes Hull (1987: 171), only to add that "some of our conceptions are not 'just' intellectual constructs". When phenomena are subject to the reign of universal natural laws (Hull, 1987; see also Brady, 1983), we seem to be dealing with something more than a mere intellectual abstraction. But this is not the point to be addressed at the present juncture. Here, we are dealing with possible 'ways of seeing', with alternative perspectives of research and observation, which both might potentially reveal some regularity of phenomena (the meaning of which will be discussed in chapter 7).

The Objectification of the Genealogical Hierarchy

The objectification of hierarchy by homology:

In 1830, the relationships of the former friends Georges Cuvier and Etienne Geoffroy Saint-Hilaire deteriorated. A great debate in the *Académie Royale des Sciences* was triggered by Geoffroy's moderate approval of a paper presented to the *Académie* by Laurencet and Meyraux on the organization of molluscs (Appel, 1987): is it possible to reduce *Sepia* and fishes to a common ground-plan of organization? Geoffroy Saint-Hilaire consented that on the basis of knowledge then available, this was a rather far-fetched idea. Nevertheless, the approach was in accordance with his own philosophical outlook, which is why he favorably reviewed the paper in question. Cuvier strongly opposed such violation of empiricism: cephalopods and vertebrates were classified by him as members of different *embranchements*, and as such it was impossible that the two should share similar structural characteristics. And indeed, the hypothesis of a *unité du type* of molluscs and vertebrates even violated Geoffroy's own criteria of "analogy" (homology in modern terminology): the nervous system lies dorsal to the digestive tract in vertebrates, but ventral in molluscs, as Cuvier succinctly pointed out.

Geoffroy was not opposed to a rational method of classification, to the formulation of a catalogue of nature which would permit information storage and retrieval with respect to the organisms included as Cuvier would have it. He also acknowledged the fact that the subordinated hierarchical classification adopted by Cuvier necessarily implied discontinuity. But he refused to see classification as an end to itself. He considered it as but a first - albeit necessary - step in the investigation of nature, producing results which have to provide the foundation for the ascent to a higher level of cognition in the science of zoology. If classification emphasizes the discreteness of groups and therewith discontinuity, a higher level of cognition, obliged to a philosophical outlook, must seek continuity in order to reduce the multiplicity of organic appearances to the philosophical concept of the unity of type (Geoffroy Saint-Hilaire, 1830: 51, 89, 121, 146; 1833a: 65). The explanation for such relations of similarity abstracting from form and function, is provided by the hypothesis of metamorphosis or transformation of essentially similar and equivalent parts of a common structural ground-plan in the actualization of different organisms during the process of ontogenetic development. This corresponds to the *transformational* conceptualization of homology.

Let us return to the problem of the homology of the opercular bone in fishes (see chapter 3). Geoffroy Saint-Hilaire understood the opercular series of bones in the fish skull to be "analogous" (homologous in modern terminology) to the auditory ossicles in mammals. This relation of similarity resulted from the *principe des connexions*, itself dependent on the *principe de l'unité de composition*: nature has at her disposal only a finite number of primordial constituent elements, defining a common structural plan, from which to build organisms. The opercular series of bones appears dissimilar as compared to the auditory ossicles of mammals, and the elements serve a different function, but this does not refute similarity based on topological relations. The same constituent elements, typical of a *"fond commun d'organisation"*, in this case of the vertebrate type (*"embranchement"*), become transformed or "metamorphosed" during ontogeny in order to warrant adaptation to different environmental conditions. In different environments, the same constituent elements take on different shapes and functions, which is why they appear dissimilar, but their reduction to equivalent constituent elements on topological grounds discloses their *essential* similarity! The *principe des connexions* allows the transformation of essentially equivalent structures in form and function, but does not permit their transposition (Brady, 1985). The metamorphosis or transformation of equivalent elements need not be revealed through observation; indeed, it cannot be observed (except in cases of ontogenetic recapitulation by terminal addition: see below), which is why topological relations (rather than specific form and/or function) must serve as guide to essential similarity. The relation of homology was, in Geoffroy's transformist view, a philosophical concept. But the hypothesis of metamorphosis or transformation explaining the relation of homology is a necessary one, because without it there is no continuity, no unity of type, but only a multiplicity of organic appearances.

Cuvier, in contrast, strongly opposed the "analogy" (homology in modern terminology) of the opercular bones in fishes with the auditory ossicles in mammals. He claimed instead that the opercular series of elements is typical for fishes only, restricted to fishes without "analogy" in any other vertebrate class. Analysis of this proposition shows it to imply two important points: notwithstanding his disagreement with Geoffroy, Cuvier's claim would seem to imply the *principe des connexions* and therewith the concept of homology, but homology was understood by him in a different sense! The statement that the opercular series of bones is restricted to fishes implies the recognition of those same elements constituting the opercular series throughout the class. This cannot be achieved without the comparison of various fish skulls with respect to the topological relations of their constituent elements, although Cuvier would have based his conjecture of similarity on functional considerations. The claim that the opercular bones are restricted to fishes implies the concept of homology of these bones throughout the group. One can go even further: only to use the term "opercular bone" means to designate a character, but to designate a character implies a relation, i.e. a comparison between minimally two organisms, preferably between three (three taxon statement: see chapter 6). It is impossible to recognize the opercular bone, its presence or absence, as a character unless a comparison is made of the thing to be characterized as opercular bone with something else that shares its presence or lacks the bone. The recognition of the opercular bone as a character of fishes thus implies the relation of homology, but homology in a different sense than understood by Geoffroy Saint-Hilaire! In Cuvier's terms, homologies diagnose groups. They do so, because specific characters serve functions which are specific for the given taxon. Beyond this teleological outlook of Cuvier, the recognition of homology generally presupposes some relation, i.e. a comparison between two, preferably three "character-bearers" or semaphoronts (see Hennig, 1967, for a definition of the term). If two semaphoronts share a homology, i.e. a character not shared by the third semaphoront, then the two first semaphoronts form a group. This is the *taxic* approach to homology: to recognize a homology is to recognize a group (Patterson, 1982), or to put it in other words: it is impossible to recognize a homology independent of some hypothesis of grouping (Rieppel, 1980).

A modern textbook of animal morphology might state that the primary jaws of gnathostomes are derived from the mandibular branchial arch of agnathan fishes. This is a transformational statement of homology: mandibular branchial arch and primary jaws are homologues, one being derived from the other by a process of transformation or metamorphosis, altering both structure and function, but not the relative position of the element in question. Similarly, the textbook might claim that the tetrapod limb, incorporating an unpaired proximal element (humerus / femur) jointed to paired distal elements (radius, ulna / tibia, fibula), is derived from a sarcopterygian fin. Again, a hypothesis of transformation is implied, explaining the change of form and function of the constituent elements which preserve their relative positions. Transformational homologies such as these imply a process of phylogenetic transformation of adult structures which is not observable (except in cases of ontogenetic recapitulation by terminal addition). This raises the danger that *a priori* assumptions about the likelihood of certain changes taking place during phylogeny might influence conjectures of homology (Rosen, Forey, Gardiner, and Patterson, 1981; Patterson, 1982). This would mean that the relations of homology, and the pattern objectified by them, could no longer be used to evaluate (test: Brady, 1985) hypotheses about process other than in a circular way.

The taxic approach to homology avoids this pitfall. Jaws are homologous throughout the Gnathostomata as indicated by their topological relations. Jaws therefore characterize a group, the Gnathostomata, i.e. recognition of the Gnathostomata as a group implies the homology of their jaws. No statement needs to be made about the mandibular arch of the primitive vertebrate skull! If on the basis of other characteristics the agnathan fishes were to be recognized as the group most closely related to the gnathostomes within a more inclusive taxon, say Vertebrata, and if this hypothesis of grouping is to be explained on the theory of descent with modification, one could then consider the possible derivation of the jaws from the mandibular arch without moving in circles: either by the addition of terminal steps of metamorphosis, or by divergent development from an equivalent ontogenetic precursor. Similarly, the tetrapod limb may be considered homologous throughout tetrapods and hence characterize the latter as a group. Recognition of the Tetrapoda as a group implies the homology of the tetrapod limb. No hypothesis of transformation is implied so far. Only if the sarcopterygian fishes are recognized as closest relatives of the tetrapods in an independent pattern analysis does it become sensible to ask the question of a derivation of the tetrapod limb from the sarcopterygian fin or of their derivation from a shared ontogenetic precursor. It might be tempting to reverse the argument: the observation of shared developmental stages in the differentiation of paired appendages by sarcopterygians and tetrapods, not shared by other gnathostomes, might be held to constitute empirical evidence for a hypothesis of hierarchical grouping. This argument amounts to phylogeny reconstruction by the ontogenetic method, presupposing the homology of shared developmental stages or rudiments and failing in the case of paedomorphosis (to be discussed in detail below; see chapter 6).

The transformational approach has been dubbed idealistic, because evolutionary transformations of structures are not observed (except by terminal addition to ontogenies which are considered to be ancestral), but rather inferred: on the basis of some hypothesis of similarity on the one hand, and on some *a priori* assumptions concerning the likelihood of a given transformation on the other. In contrast, the taxic approach has been claimed to be empirical (Brady, 1985), resulting in the discovery or recognition of natural groups which are "real", i.e. which "exist outside the human mind" (Patterson, 1982) and therefore exist irrespective of our willingness or ability to discover or recognize them. This justification of the taxic as opposed to the transformational approach is a little too enthusiastic, however, since both 'ways of seeing' depend on the test of similarity, and hence presuppose methodological rules (see above). Homologies must minimally pass the test of similarity, i.e. the constituent elements to be homologized must be in a comparable topographical relation to each other and with respect to a fixed frame of reference, if the question of evolutionary transformation is to be meaningfully addressed at all. Alternatively, the features supposed to diagnose a group, whether shared developmental stages or adult characters, must likewise pass the test of similarity, since they, too, vary in shape and function and since the conjecture of taxic homology on grounds of congruence alone is circular. The claim that the Gnathostomata are characterized by the presence of jaws presupposes the homology of the jaws throughout the group, irrespective of their manifold differentiation in form and function. In a similar vein, the recognition of the Tetrapoda implies the homology of the tetrapod limb in animals as different as frogs, birds, bats, horses and seals. The limbs in these subgroups of the Tetrapoda vary extensively in form and function, so that their homology must be based on topological relations, while at the same time it must be acknowledged that the limb can become transformed into a variety of shapes to serve a variety of functions. A 'compositional' criterion may be derived from the internal organization of the tetrapod limb, viz. its construction from an unpaired proximal element, articulating with paired distal elements which in turn relate to the autopodium. But again, these 'compositional' criteria denote nothing but topological relations at a less inclusive hierarchical level than "paired appendages", i.e. at the level of the Tetrapoda which is subordinated to the level of the Gnathostomata.

Patterson (1982: 38) has claimed that transformational homologies fail the test of similarity, illustrating his point with reference to Bjerring's (1977) conjecture of homology of the basicranial muscle of *Latimeria*, assumed to exist in fossil "crossopterygians" too, with the polar cartilages of other vertebrates. "Crossopterygians" (Actinistia or coelacanths, of which *Latimeria* is the only living representative, and the entirely extinct porolepiform and osteolepiform rhipidistians) are unique among vertebrates in the shared presence of a functional intracranial joint. There still persists some disagreement as to the primitive or derived status of this joint which subdivides the neurocranium and the dermal skull roof into an anterior and posterior portion, the snout complex being hinged on the posterior division. In *Latimeria*, a basicranial or subcephalic muscle was found to bridge this joint ventrally, effecting its flexion upon contraction. A functional analysis of the coelacanth head was provided *inter alia* by Thomson (1966, 1967, 1970) and Lauder (1980).

The polar cartilages, on the other hand, have been identified in a number of vertebrates as the posterior portions of the trabeculae cranii, linking the latter to the parachordal basal plate and giving rise to the laterally projecting basitrabecular processes. In short, the polar cartilages are elements of the embryonic basicranium, lying in a position which corresponds to that of the ventral portion of the intracranial joint in "crossopterygian" fishes.

The homology of the structures in question is rejected by Patterson (1982: 38) on the grounds that "the compositional criterion is denied" and "the ontogeny of the muscle is unknown". What exactly is to be understood by the "compositional criterion" in the present context is difficult to comprehend. The 'compositional criterion' may be just another way to express topological relationships, as in the case of the composition of the tetrapod limb. If understood in that sense, Bjerring's conjecture of homology does not fail the test, because he bases his hypothesis of similarity on the observation of similar topological relations of the structures in question. He derives both, the basicranial muscle as well as the polar cartilage from the mesoderm of the third (mandibular) head segment: if this topological argument is to be refuted, criticism must be aimed at Bjerring's (1977) elaborate theory of head segmentation. Alternatively, his conjecture of homology is obviously a transformational one, as it implies the possibility that cells which form muscle tissue in some vertebrates, give rise to cartilage in others (Bjerring, 1977: 174). The hypothesis of homology may be thought to fail the compositional criterion, because the structures in question are composed of muscle cells in some, of cartilage cells in others. As was discussed in chapter 3, the transformational relation of homology reduces varying organic appearances to originally equivalent precursors, constituent elements of the type. If Bjerring's hypothesis is to be refuted on these grounds, it must be shown that precursors giving rise to muscle tissue cannot also give rise to cartilage. This, however, cannot be demonstrated! On the contrary, experiments on limb regeneration in amphibians have shown that dedifferentiated muscle cells can give rise to the limb skeleton in the regenerate (see Hall, 1978: 115-119, for further details). Assuming an equivalent and reversible potency of embryonic precursor cells to form any of the three, muscle tissue, cartilage, or bone, there is no reason to reject the homology of the subcephalic muscle and of polar cartilages on compositional grounds. It is true, however, as argued by Patterson (1982), that the ontogeny of the basicranial muscle is not known. However, this is also true for many structures for which conjectures of homology have been proposed, and the delamination versus invagination of the neural tube in vertebrate embryos shows that development by itself is no infallible guide to homology of topologically equivalent adult structures.

Nevertheless, Patterson's (1982) rejection of Bjerring's hypothesis on grounds of ontogeny is justified, not because the ontogeny of the basicranial muscle is unknown, but because the homology of that muscle with polar cartilages critically depends on Bjerring's (1977) model of head segmentation. There is no doubt that this model is far-fetched and based on a number of conjectures which are simply beyond test. The homology of the subcephalic muscle with polar cartilages stands or fails with the theory of head segmentation as adopted by Bjerring (1977), but criticism and eventual refutation of that theory will necessarily have to be based on topological considerations.

Transformational homologies may deny the test of conjunction by definition (Patterson, 1982: 38). Bjerring's hypothesis of a transformational homology would be refuted if an animal were found which combines the presence of a subcephalic muscle with polar cartilages. Classical hypotheses of homology denying the test of conjunction are serial homologies *sensu* Owen (Russell, 1916: 109). Zangerl and Case (1976) have recently resurrected Gegenbaur's theory according to which the pectoral and pelvic girdles and the limb skeleton are homologous with branchial arches and gill rays. As Patterson

(1982: 35, 38) noted, this relation of homology is not refuted by the fact that a gnathostome has both structures. Similarly, the test of conjunction may be denied by ontogenetic transformation of structures in the course of the development of the same individual organism. An example is provided by the ear ossicles of mammals. All tetrapods except mammals are characterized by a single stapedial bone, distally in contact with the cartilaginous extracolumella (if present). Mammals alone among tetrapods are characterized by the presence of three auditory ossicles, viz. malleus, incus and stapes. The presence of three auditory ossicles diagnoses a group within extant Tetrapoda, i.e. the Mammalia. No hypothesis of transformation is implied so far.

The theory of C.B. Reichert, developed in 1837, specifies that the articular bone of non-mammalian vertebrates is to be homologized with the mammalian malleus. The primary jaw articulation (between articular and quadrate bones) of non-mammalian tetrapods is replaced in mammals by a secondary jaw articulation (between dentary and squamosal), while the articular bone is incorporated into the sound-transmitting apparatus. Within the series of ear ossicles, the articular (malleus) retains its original topological relations, as it continues to articulate with the quadrate which itself has become the incus lying proximal to the malleus, between the latter and the stapes. In other words, the primary jaw articulation of non-mammalian tetrapods is preserved within the chain of ear-ossicles of mammals. The whole theory of Reichert deals with idealistic transformations, but it has been supported by reference to some marsupials which retain a primary jaw-articulation during the first three weeks after birth. The replacement of the primary by a secondary jaw articulation thus takes place in a postnatal phase of development (Starck, 1979: 341). The argument presupposes the correctness of Reichert's theory, i.e. it rests on the acceptance of the conjectured homologies, but accepting these for the sake of the argument, it is readily perceived that the sequential occurrence of an articular and a malleus during the ontogeny of some marsupials does not refute their conjectured homology: the test of conjunction is, again, denied by definition.

Both the taxic as well as the transfomational conceptualization of homology must pass the test of similarity: if that is the meaning of the predicate 'idealistic', then both types of homology must bear that predicate, since both types of homology are conjectured on the basis of the operational *principe des connexions*. Homology, whether in the taxic or in the transformational sense, whether of developmental stages or adult structures, denotes a logical relation of form as it emerges from a methodological rule. This relation to a common background indicates a basic comparability of taxic and transformational homology. Transformational homologies, and in particular conjectures of serial homology or homonomy, deny the test of conjunction by definition. Taxic homologies may fail this test, too, if they form part of a recapitulated ontogenetic sequence. In one sense, however, taxic homologies are more rigidly testable than transformational homologies. Since the latter imply no hypothesis of grouping, but only a sequence of evolutionary transformations, the test of congruence is not applicable. The transformation of the mandibular arch into jaws is not testable by congruence, but the hypothesis of the Gnathostomata as a subgroup of the Vertebrata can be subjected to this test.

Both types of homologies have to pass the test of similarity; both types may deny the test of conjunction; but only taxic homologies can be tested by congruence. One of the relations of the taxic and transformational homologies, therefore, is that the first type is more rigidly testable. If testability and potential falsifiability are a measure of empirical content, it follows that the empirical content of taxic homologies is higher than that of transformational ones.

The other relation between the taxic and the transformational approach to the concept of homology is that of pattern versus process. The taxic approach to homologies focuses on discontinuity, specifying a subordinated hierarchy of groups within groups: the presence of jaws, irrespective of their form and specific mode of function, characterizes the Gnathostomata within the Vertebrata; limbs characterize the Tetrapoda within the Gnathostomata. The transformational approach on the other hand seeks to understand how organs, or organ complexes became transformed into one another - during ontogeny and phylogeny. However, only the transformations of those characters which appear in the hierarchy of homologies as revealed by the taxic approach can at all be meaningfully addressed. Only if it is agreed that 'paired appendages' are a character, and therewith a homology, of the gnathostomes is it possible to ask whether the differentiation of these 'paired appendages' in fishes as opposed to vertebrates, i.e. in different subgroups of the gnathostomes, can be meaningfully addressed in the context of a transformational perspective. The question whether the tetrapod limb is more similar to a "crossopterygian" or a dipnoan fin (Rosen, Forey, Gardiner and Patterson, 1981) bears a number of hidden implications: a) both the sarcopterygians ("crossopterygians" plus dipnoans) and the tetrapods are gnathostomes, b) therefore, their fins and limbs are different appearances (metamorphoses) of the gnathostome character 'paired appendages, c) the sarcopterygians are the closest relatives of the Tetrapoda within a more inclusive taxon, itself subordinated to the Gnathostomata. Given that hierarchy of groups within groups, and the homology of 'paired appendages' throughout the Gnathostomata, the call for transformation as an explanatory hypothesis for the observed pattern becomes meaningful. Riedl (1975, see also Rieppel, 1983a) used the terms "*Rahmenhomologie*" ("frame-homology") and "*Spezialhomologie*" ("special homology") to denote the hierarchical relation of taxic versus transformational homologies. The character 'paired appendages' is the *Rahmenhomologie*, characterizing the Gnathostomata; the sarcopterygians and the tetrapods are two subgroups of the gnathostomata, characterized by a special (in the sense of von Baer, 1928, meaning less widely distributed) differentiation of the paired appendages, and the transformational approach addresses the question how these "special homologies" are connected with each other within the more inclusive concept of the general homology (for a more detailed discussion see the following chapter).

On this basis it becomes obvious that the transformational approach to homology, if it is to be meaningful, must presuppose taxic relations, based on character analysis or taxic homologies. This is why both types of homologies are related to each other in the sense that both must pass the test of similarity in order to be acceptable. That the taxic approach to homology should have precedence over the transformational approach is not only a consequence of logical and philosophical relations (Brady, 1985, and above), but is furthermore justified by the higher empirical content of the taxic relation of homology, testable by congruence. The transformational approach serves to bridge the gaps separating taxa by a hypothesis of process, i.e. of descent with modification.

If, on the basis of a foregoing pattern analysis, the question is raised whether jaws could be derived from the mandibular arch or the tetrapod limb from a sarcopterygian fin, the taxic approach is implicitly abandoned and the analysis becomes one of transformations. Problems of evolutionary transformations entail no hypothesis of subordinated grouping, but a hypothesis of graded similarity mirrored by the serial arrangement of organisms in a continuous morphocline. As was emphasized by von Baer (1828) already, the hierarchy of types denotes *relations of coexistence* of constituent elements in organic structures (Rieppel, 1985b). No transformation of one type into another is implied: the tetrapod is always a tetrapod, throughout its development, although it shares with fishes early developmental stages of an "imperfect" vertebrate. The taxic approach results in a hierarchy of fixed types representing logical relations of form, and in that sense it is ideal as well as static. The transformational approach, however, puts the hierarchy of homologies into a temporal sequence of form in order to ask the question of evolutionary transformation. The static hierarchy of types is reduced to a temporal sequence of forms: homology comes to denote *relations of succession*. The direction of evolutionary transformation must be inferred from the polarization of the morphoclinal sequence of graded similarity as it results from the assessment of the generality of character distribution: evolutionary transformation is inferred to proceed from the more general to the less general, i.e. along the axis of the cladogram depicting the hierarchy of (taxic) homologies. This interplay between taxic and hierarchic approach reveals a basic aspect of the complementarity of subordinative versus serial hierarchy, of pattern versus process, a relation which will have to be dealt with in greater detail in the next chapter. Pattern analysis, i.e. the taxic approach, must precede the explanation of pattern by process; at the same time, it provides the hypothesis of process with a direction. On the other hand, the hypothesis of process provides pattern with a causal explanation.

The objectification of hierarchy by ontogeny:

It was claimed above that the genealogical hierarchy is mirrored by the hierarchy of homologies, and that the ontogenetic process of transformation or differentiation runs parallel to the hierarchy of homologies, proceeding from the more general to the less general. The corollary is that the genealogical hierarchy can not only be objectified by the relation of homology, but also by ontogeny. The ontogenetic sequence of developmental stages would provide a key to the phylogenetic past.

As in the case of homology, it is important to distinguish the transformational and the taxic approach with respect to the ontogenetic process of differentiation, too. It was shown above that ontogeny can be viewed in the light of an atomistic or an epigenetic paradigm. The atomistic paradigm emphasizes the continuity of transition by the mechanism of terminal addition. It was outlined above that the mechanism of (terminal) addition is related to the Aristotelian concept of a *scala naturae* composed of a series of privative taxa (except the terminal one) which, on that mechanism, can be linked up with each other in a continuous series of ancestors and descendants. It is obvious that the concept of terminal addition can most readily be interpreted as a mechanism giving rise to a serial sequence of ancestral and descendant taxa. As was noted (chapter 4) with reference to the insertion of the amnion into an ancestral ontogeny, however, such non-terminal addition does not change the topography of the hierarchy. It still allows the translation of a segment of a branching hierarchy into a series of ancestors and descendants. The most crucial point is that atomism conceptualizes phylogeny as an essentially additive, i.e. cumulative process. The epigenetic 'way of seeing', on the other hand, emphasizes the subordinated hierarchy and hence the discontinuity of the appearance of organic beings as follows from ontogenetic deviation leading to the individuation of the terminal taxa. At the same time, atomism is reductionistic whereas epigenesis has a more holistic background.

It has been detailed in the preceding chapters that atomism views organization as a result of the juxtaposition of parts or constituent elements, and that it is on this perception that the theory of evolutionary transformation was based. For Darwin, variation resulted from the fortuitous aggregation of parts derived from both parental bodies in the generation of offspring. Evolution on the basis of variation and differential sorting was viewed by him as an additive process, allowing the adaptation of organisms to changing environments over time (Schweber, 1985: 49). This view of life was embedded in Darwin's coral reef theory: the reef grows continuously in response to changing environmental conditions, producing an uninterrupted series of forms (Gruber, 1985: 19). This rather archaic theory has been replaced in neo-Darwinism by the re-discovery of the Mendelian mechanisms of heredity, variation now being explained by the fortuitous recombination of genes derived from both parents. Particularly the early attempts to incorporate mechanisms of genetics into a theory of transformation show the influence of an atomistic background. This is clearly borne out by Mayr's (1982) treatment of the mutation theory of Hugo de Vries as related to the work of Gregor Mendel.

Mayr (1982: 713-714) characterized Mendel's theory of inheritance, "stripped to its essentials", as a hypothesis requiring "that for each heritable trait, a plant is able to produce two kinds of egg cell and two kinds of pollen grain, each representing either the paternal or the maternal character", which is the same as saying that each trait is "represented in the fertilized egg by two hereditary elements (and no more than two), one derived from the mother...and one from the father...". In Hugo de Vries' views, these hereditary elements turned into genetic units ("pangen") each coding for a certain trait and varying independently from one another; should a genetic unit mutate, the stage would be set for the development of a new variety or species (Mayr,1982: 707-709). A quotation of de Vries given by Mayr (1982: 707) reveals the atomistic background in a particularly striking manner:

> "But if the species characters are regarded in the light of the theory of descent it soon becomes evident that they are composed of single factors more or less independent from each other"

In the modern, synthetic theory of evolution, the atomistic view of variation and hence of organization has become couched in the statistical language of population genetics (Webster and Goodwin, 1982). It has been recognized that "beanbag genetics" (Mayr, 1963: 263) is a misleading concept because of the interaction between hereditary components at various hierarchical levels of genotype and gene-pool organization, but the origin of new species is still believed to result from a reconstitution of the gene-pool of a founder population (Mayr, 1963: 527). A "bag of beans" determines, by its interaction, a change in a character complex, i.e. in the phenotypic combination of characters (by addition or substraction) or character-states (by mutation causing metamorphosis of an existing character). The reductionist tendency of the atomistic approach to transformation is revealed by the tendency to explain large-scale evolutionary events (macroevolution), i.e. the origin of new types of organization, by mechanism of microevolutionary change as they take place within populations: "Indeed, does not all the available evidence suggest that the difference between minor and major types are merely matters of degree?" (Mayr, 1963: 589). Even to ask the question of the origin of a new *type* of organization reveals, in the atomistic perspective, a typological or essentialistic thinking to be abandoned because "the gene-frequency approach of population genetics was quite unable to supply any solution to this problem of origination" (Mayr, 1982: 616).

Variation of the genotype is postulated to be fundamental, and so is the variability of the phenotype, potentially affecting the organism in all its parts. Fundamental variation renders the gradual, stepwise transformation of organisms possible: they may be affected by slight changes in any and all of their parts. Theoretically at least, all possible intermediates between two complex structures must obtain, as Darwin emphasized, and as is still claimed by protagonists of the synthetic theory of evolution (Bock, 1979). Transformation by terminal addition, advocated by Meckel and Serres, implies terminal addition of new parts, or of new steps of metamorphosis of existing parts, to an ancestral ontogeny. Whether terminal or not, the result of an additive process is the gradual transformation of form, graphically represented by a linear hierarchy or serial arrangement of organisms. Transformation according to this atomistic model proceeds stepwise, which is why "...numberless intermediate varieties linking most closely all species of the same group together, must assuredly have existed..." (Darwin 1859 [1959:153]). All of the successive steps of transformation are tested by natural selection, from whence selection derives its "creative power": Mayr (1963: 201) compared the action of natural selection with that of a sculptor, chipping off bits and pieces from a block of marble to produce a

113

piece of artwork. Each step of change or metamorphosis, however small it may be, is tested, and either admitted or discarded. Natural selection assumes a guiding role in the process of transformation, ensuring the adaptiveness of evolutionary change. The smaller the evolutionary innovation subjected to test is, the more easily it can be discarded or modified in an adaptive direction. The more trenchant the effect an evolutionary change has on the phenotype, the greater the risk of a misfit in relation to environmental conditions, and the more destructive will the action of selection be, relegated to a purely eliminating function.

The creative power of natural selection, guiding evolution along the road of functional adaptation, must be based on a finely graded series of transformational steps, a process requiring extended periods of time: "Adaptive evolutionary change requires strong directional selection acting over a long period of time" (Bock, 1979: 54) in order to effect the shift from one "adaptive zone" to the other. New types of organization originate through a succession of intermediates; different types of organization are viewed as adaptations to different types of environment. Evolutionary change, whether reduced to a shift of gene frequency within a population or whether extrapolated therefrom to explain the origin of new types by the invasion of new adaptive zones, is causally determined throughout. Causal determination as understood in this context refers to the determination of any structure by its function, to the shaping of the organism according to environmental requirements by the force of natural selection. Pervading causal determination of form in relation to function requires the continuous action of natural selection, "daily and hourly scrutinizing, throughout the world, every variation, even the slightest; rejecting all that is bad, preserving and adding up all that is good" (Darwin, 1859[1959: 72]). Pervading causal determination of evolutionary change requires acceptance of "that old canon in natural history of 'Natura non facit saltum'", as Darwin (1859 [1959: 166]) put it.

Behavior is viewed to play a major role in the initiation of evolutionary change (Mayr, 1963: 605): "...changes in behavior are indeed considered important pacemakers in evolutionary change" (Mayr, 1982: 611), "...many if not most acquisitions of new structures in the course of evolution can be ascribed to selection forces exerted by newly acquired behavior" (Mayr, 1982: 612; see also Mayr, 1974). An example is provided by Bock's study of the evolution of foot structure in woodpeckers; the new behavior of climbing trees set up new selection pressures which modified the ancestral foot structure still retained in what is interpreted as representing primitive species of the group. Evolution proceeds in a stepwise fashion according to the new selectional regime by what has been called "evolutionary tinkering" (Jacob, 1983), recombining parts, adding new, substracting old or modifying those already existing.

The functionalist 'way of seeing' (Lambert and Hughes, 1984; Hughes and Lambert, 1984) conceptualizes natural selection as the creative force, i.e. the efficient cause of transformation. Function, tested by natural selection, determines structure. Since all structures are differentiated or created in the course of ontogenetic development, functionalists are forced to postulate that ontogeny is molded by natural selection. While this cannot be denied in the case of larval or embryonic adaptations, it is hard to see how the tax of selection, imposed on the sexually mature phenotype, should affect the differentiation of this phenotype during its embryonic phases of development. Such can only be visualized if evolutionary transformation is conceptualizd to result from the gradual (step-wise) modification of terminal stages of ancestral ontogenies which, at one time in the phylogenetic past, had been evaluated in terms of Darwinian fitness, i.e. reproductive success.

Functionalism is opposed by the epigenetic view of development which is obliged to a more holistic view of organization, and thus to the structuralist 'way of seeing' (Hughes and Lambert, 1984). Etienne Geoffroy Saint-Hilaire was asking for general laws of form, and he found these reflected in his *principe des connexions*. This principle secured the primacy of structure over function. Environmental conditions were allowed to exert a modifying effect on the developmental process, but only to an extent that would not disrupt the basic organization of the type. In his discussion of vertebrate digit patterns, Holder (1983: 452) noted that "if internal constraints on morphological variation did not exist, it may be expected that specific morphological characteristics would vary extensively at the whim of external environmental influences". But "although many anatomical variations occur during the evolution of the limb skeleton, they all occur within a clear anatomical framework", or "according to a basic and identifiable plan" (Holder, 1983: 451). Geoffroy explained the constancy of structure on the hypothesis of the *nisus formativus*, and therewith qualifies as a vitalist from the modern perspective. This does not imply, however, that the concept of the type, i.e. of a common ground-plan of organization must be abandoned in modern work; rather, the permanence of a given *bauplan* may be explained on the hypothesis of developmental constraints. Shubin and Alberch (1986: 377) write:

> "...this typological [essentialistic] and static approach is not opposed to evolution or even to natural selection. It emphasizes invariance over variation and regularity and constraints over chance and contingency...The quest for a general set of principle of form is legitimate if we exchange the metaphysical concept of the *Bauplan* for a mechanistic one based on principles of morphogenesis and internal integration."

There is no need to continue to evoke transcendental laws of form or forces of development. Instead, the structural 'way of seeing' emphasizes "stability of structure, the difficulty of its transformation" (Gould, 1982a: 90), and the regularity of the hierarchy of groups within groups which follows. The question then is: why should classification be at all possible, why does the diversity of form take on an ordered appearance? After all, writes Gould (1983: 362), there is no need to disregard the study of internal sources of stability while accepting externalist explanations as all-embracing just because of an "exaggerated fear of essentialism".

> "For if variation is fundamental and species have no meaningful essence, then what can hold a species in one region of morphology except an external force that culls and regulates its defining variation? Developmental programs and genetic systems are 'essential' properties 'distributed' to individuals by virtue of their membership in a species...The admission of such essential properties might force evolutionists...to consider the role of 'internal factors' in evolutionary change and stability" (Gould, 1983: 362-363).

Viewed from this perspective, form and its transformations do not move continuously through morphospace, but rather concentrate on some basic types of construction or organization, "stability domains" (Oster and Alberch, 1982: 455; see also Alberch,1980) which correspond to the hierarchy of groups within groups. Natural selection loses its omnipotence to mold structure and function in any direction required by environmental conditions, even beyond the limits of a given structural plan, creating new structural plans in response to new selective pressures. Selection may optimize adaptation within the limits of a given type, but from the structuralist point of view, it cannot create a new type from an ancestral one. The type is determined by developmental constraints, or in other words: a structure may provide a selective advantage, but its function does not, or not sufficiently, explain its origin.

Viewed from a structuralist perspective, ontogeny is conceptualized as a dynamic and creative process. Development corresponds to the actualization of potentialities for structural differentiation. The potential for divergent development is restricted or canalized by regulative mechanisms governing the process of ontogeny. A new type is created by ontogeny, i.e. by a change of generative mechanisms of embryogenesis. Alberch and Gale (1985) have investigated digital reduction in amphibians, comparing patterns of phylogenetic reduction to experimentally induced reduction. Based on mechanistic grounds a number of possible morphologies had been predicted which were found to match the natural occurrences of form to a high degree. However, some "differences between expectations from ontogeny and the morphologies observed in nature" seemed to "imply functional (selective) constraints on design" (Alberch and Gale, 1985: 19). Selection can have an effect on form, but only within developmentally constrained types. The type, or rather the principles of form on which it is based, determines the options of adaptation: structure determines function. And since structures are created by ontogeny, it must follow that from a structuralist point of view, ontogeny makes evolution rather than being molded by evolution. Morphological evolution is then "defined as the change in developmental programs during phylogeny" (Oster and Alberch, 1982: 444). The development creates a morphology which, once formed, is submitted to the test of natural selection - and either rejected or maintained and perhaps improved as far as possible. As clearly perceived by Mayr (1982: 201), the 'typological' or 'essentialistic' way of seeing imposes on selection an eliminating rather than a creative power:

> "We need to view the organism as an integrated whole, the product of a developmental program and constrained by developmental and functional interactions. In evolution, selection may decide the winner of a given game but development non-randomly defines the players" (Alberch, 1980: 665).

Sober (1984) discusses the issue of adaptation at some length. How is the origin of a new structural characteristic to be explained? "A characteristic that initially is present fortuitously can be selected and thereby become established in a population. Initially, it is not an adaptation for doing anything; later on, it is", writes Sober (1974: 157). Beyond mere accident, structuralism holds that the initial appearance of a new characteristic must ultimately be due to the creative powers of ontogeny; later on, it may become modified and adapted by the action of natural selection.

It may be noted in this context that modern structuralism achieves the synonymy of the two historically consecutive meanings of the notion of 'evolution'. Charles Bonnet introduced the term *évolution* into the field of natural history in order to denote the development (unfolding) of pre-existent germs. His view corresponded to what Piaget (1967 [1974: 134]) characterized as a "structuralism without genesis". All types of organization were preformed at the time of Creation, and even if those types became actualized during different epochs of the earth history, no transformation, i.e. no genesis of a genuinely new type, is implied. Following the work of Lamarck, and especially of Charles Darwin, the "structuralism without genesis" gave way to a "genetism without structure" (Piaget, 1967 [1974: 134]). Organisms became parts of an ever-changing world, being subject to continuous transformation. Types of organization continuously blend into each other over geological time, so that all demarcation of clear-cut types of organization must result from the abstracting powers of ordering intelligence rather than being founded in nature.

"Structure seen through a developmental perspective is not only descriptive but also central to the understanding of change in form, i.e. evolution" (Lambert and Hughes, 1984: 478). Looking at this sentence in isolation, and replacing the word 'development' for its

French synonym *évolution*, it must be acknowledged that the meaning of this term is utterly ambiguous. *Evolution* may be read as to refer to the process of embryonic development which consists of nothing but a series of transformations or metamorphoses, creating new structures in a divergent process of differentiation, or adding new modifications to existing ontogenies. *Evolution* may also be read as to refer to the process of phylogenetic development which in effect is nothing but the outcome of changing ontogenies. The synonymy of these two meanings represents the basis for the claim that ontogeny provides the only empirical (observational) access to an understanding of transformations, and that the order of ontogenetic transformations is expected to be reflected in the order of nature which is interpreted as the result of the phylogenetic process (Nelson, 1978; Patterson, 1982).

The fact that structures or types of organization are not distributed all across the morphological continuum supports the view that evolution, in both senses of the word, must be constrained by regulative mechanisms inherent in structure. What is at issue then is a "genetic structuralism" (Piaget, 1967 [1974: 134]), striving towards an understanding of the constancy of form as much as the possibility of its constrained transformations. The basic question, lying at the heart of the structuralist approach, must therefore address developmental mechanisms which make certain morphologies possible while "forbidding" others (Stock and Bryant, 1981). The epigenetic process creates form through a process of growth, compartmentalization and differentiation starting out from a homogeneous primordium. The resulting organism is a developmentally and functionally integrated whole, a harmoniously structured *"Tout organique"* in Bonnet's terminology. In this context it is interesting to recall a quotation taken from Bonnet and compare it with one taken from a modern structuralist source, keeping aware of the rather close connections of Bonnet's preformationism with Harvey's theory of epigenesis (Rieppel, 1985c, 1986a), particularly at the more advanced stages of development of Bonnet's theory of pre-existence: "...this term [viz. the pre-existent germ] will not only designate an organized body *reduced to small size,* but it will also designate all kinds of *original preformation, from which a Tout organique can result as from its immediate cause"* (Bonnet, 1769, vol. I: 362). On the basis of the principle of continuity, Bonnet (1768, vol. I: 4-5) asserted: "The actual state of an organism is the consequence or the product of the preceding state, or to put it more correctly, the present state of an organism is determined by the preceding state". This does not sound terribly different from Oster and Alberch's (1982: 450) claim that

> "While the stage definitions are arbitrary temporal landmarks in a continuously unfolding [note that this is the original meaning of the word *évolution* as coined by Bonnet!] process, nevertheless it remains true that the conditions characterizing the end of one stage set the initial conditions from which the next stage must proceed. That is, the sequential nature of the geometrical changes characterizing morphogenesis and differentiation provide the most basic kind of constraint: each stage must proceed from where the last left off".

There follows, from the structuralist 'way of seeing', a continuity of form-generating causes constraining the developmental process to a limited array of possible developmental pathways which result in discontinuous phenotypical effects, i.e. morphologies restricted to certain domains of stability which are hierarchically structured in a subordinated sense. Any deviation or bifurcation of the developmental pathway is liable to create a new form, and the earlier during ontogenesis this deviation takes place, the more profound is the morphological change liable to be (Arthur, 1984). This is simply a function of the fact that any given developmental step constrains the options open to subsequent development, so that the constraining effect of a deviation will be more profound if it occurs early during ontogeny. Raff and Kaufman (1983) give an example of an early ontogenetic bifurcation creating discontinuity in nature which renders the subdivision of the Metazoa into protostomes and deuterostomes possible. The majority of protostomes are characterized by spiral cleavage of the egg cell, while deuterostomes show radial cleavage. These two basically different types of development result from nothing more than "the timing and placement of the successive mitotic spindles that are reflections of the organization of the cytoskeletal matrix of the egg" (Raff and Kaufman, 1983: 105). An important corollary is that form generation during ontogeny is not random, but causally determined throughout by the continuity of the developmental process. However, since form has primacy over structure, the creation of form during ontogeny is random with respect to environmental condition and hence with respect to the selection regime.

Another corollary is that because of the discontinuity of morphological effects resulting from the bifurcation of developmental pathways, the search for intermediate stages of differentiation is rendered futile. Oster and Alberch (1982) discuss the development of epidermal outgrowths such as hair or feathers. Invagination of the epidermal placode precedes the development of hair, while feathers result from an evagination of the epidermis. *Tertium non datur*: this classical case of a bifurcation of developmental pathways renders the existence of intermediate differentiations morphogenetically impossible.

Various models have been proposed to account for ontogenetic deviations resulting in the creation of new morphologies. Oster and Alberch (1982: 455) find morphologies to be constrained to "major themes", i.e. to "stability domains" which correspond to the concept of a type or *bauplan*. Mutation and selection are not considered powerful enough to transcend the boundary conditions of these patterns of organization, but can only act to diversify the underlying pattern without major qualitative changes. However, genetic variability may accumulate and thus reach a threshold, bringing the developmental system to a "bifurcation boundary", eventually pushing it across that boundary into a different developmental pathway: "the transitions between different morphologies are constrained by the biology of the bifurcation set..." (Oster and Alberch, 1982: 455). This model has recourse to the concept of an epigenetic landscape as developed by C.H. Waddington: the developing organism is compared to a billiard ball rolling down a sloping system of bifurcating valleys representing possible developmental pathways. These developmental pathways are separated by impossible morphologies, the impossibility of their actualization resulting from the biological properties of the control system programming the developmental process and from physical constraints imposed by the properties of the material used to construct living systems. The threshold model of developmental bifurcations derives from mathematical considerations which indicate that if a stable system reaches a critical point, a quantitatively small change can trigger a large qualitative change of pattern (Shubin and Alberch, 1986: 329).

A quite different and most interesting approach to the analysis of form has been proposed by B.C. Goodwin who advocates the change from an evolutionary to a generative paradigm in comparative biology (Goodwin, 1984a). His "relational or field theory of reproduction" (Goodwin, 1984b) relates primarily to the insufficiency or, in this author's view, inadequacy of the atomistic perspective incorporated into Darwinian evolution, and it is perhaps best introduced by a couple of examples.

Darwinism, it will be recalled, postulates the gradual transition form one type of organization to the other, from one adaptive zone to the other, and seeks to document this change by a series of intermediates. The Darwinian view of life appeared to be strikingly corroborated by the discovery of *Archaeopteryx* in the Upper Jurassic lithographic limestone of Solnhofen in the 1860s (Ostrom, 1985a). Here was a fossil which provided an almost perfect intermediate between two vertebrate classes, reptiles and birds. A thorough re-investigation of the fossil from a neo-Darwinian perspective led to the conclusion that "the immediate ancestor of *Archaeopteryx* was a small coelurosaurian dinosaur and that the phylogeny of avian ancestry was: Pseudosuchia - Coelurosauria - *Archaeopteryx* - higher birds" (Ostrom, 1976: 174). One of the many characters shared by the small and agile (bipedal) coelurosaurian dinosaurs and *Archaeopterxy* (and, by implication, higher birds) is the reduction of the manus to three digits. The problem only is to know which of the digits have been preserved in coelurosaurs, *Archaeopteryx*, and higher birds!

One way to identify digits is by the phalangeal count, which, for the primitive reptilian hand is 2-3-4-5-3. Theropod dinosaurs, to which coelurosaurs belong as an included subgroup, are commonly believed to have retained the digits I, II, and III, whereas the digits IV and V have become reduced. This interpretation is based on the phalangeal formula (2-3-4) of the preserved digits, but also on the fact that the first digit may to some degree be opposed to the other two, and that rudiments of a fourth digit may still be retained, *Ornitholestes* being an example. Accepting the above hypothesis of descent, *Archaeopteryx* and higher birds would, by implication, also have lost digits IV and V.

This is where problems begin. Recent embryological investigations of the developmental pattern of the bird wing have shown that higher birds are best interpreted to have lost digits I and V, while retaining digits II, III and IV (Hinchliffe and Hecht, 1984; Hinchliffe, 1985). If that interpretation is accepted, birds cannot be descended from theropod (i.e. coelurosaurian) dinosaurs. By implication, *Archaeopteryx* must also have lost digits I and V in order to retain its ancestral position relative to birds, or it has retained digits I, II, and III in order to be descended from coelurosaurs - but then it cannot have given rise to higher birds.

The problem might be solved with recourse to the phalangeal formula of *Archaeopteryx*. However, the preservation of the fossil results in some ambiguity with respect to the phalangeal count, whether it be 2-3-3 (Hecht and Tarsitano, 1983) or 2-3-4 (Ostrom, 1985b). Whichever count is accepted, the assumed reduction of digits and phalanges must be compatible with known morphogenetic mechanisms controlling the differentiation of the limb-bud in extant tetrapods (Hinchliffe and Hecht, 1984; Hecht, 1985).The resolution of the controversy by the proposal of a most parsimonious hypothesis for the placement of *Archaeopteryx* (Gauthier and Padian, 1985) cannot be the object of the present account: all this example is intended to demonstate is the atomistic background of the analysis: the whole argument revolves around the number of phalanges and digits. This presupposes, however, a genetic and positional identity of the digits and the phalanges of which the latter are composed, imparting (material and historical) individuality and hence identifiability on the constituent elements of the manus. This is precisely the point to which Goodwin (1984a: 113) would object:

> "Organisms are not aggregates of elements, whether molecules, cells, organs, skeletal or other components, whose random variation results in an unconstrained variety of forms. They are self-organizing wholes governed by laws describing spatial and temporal organization such that processes of biological change involve constrained transformation, whether ontogenetic or phylogenetic."

In other words: the problem addressed above is a real problem only within a certain conception of form and organization, of development, and phylogenetic transformation; it dissolves if atomism is rejected. Instead, Goodwin and his collaborators view the whole limb "as an integrated developmental field in which the pattern is defined by a small set of developmental rules" (Shubin and Alberch, 1986: 325).

The generative paradim or field theory of development is linked to an epigenetic background, and therewith to the principle of continuity: any stage of differentiation is the consequence of - or constrained by - the preceding steps of development, the constituent elements originating one from the other in a particular developmental, functional and spatial relation to each other (see Goodwin and Trainor, 1983: Fig. 9, for an illustration of that point). One characterization of the field as given by Goodwin (1984b: 228) reads : "A field is a spatial domain in which every part has a state determined by the neighbouring parts so that the whole has a specific relational structure", another definition is (Goodwin, 1984a: 107): "A field is a domain of tissue capable of forming a structure, such as a limb, with the capacity of responding as a unitary, self-organizing whole to a variety of disturbances". Such disturbances can be experimentally induced, and the observation of regular regeneration of limbs demonstrates the relative stability of the field which guarantees a restitution of the normal spatial pattern (Goodwin, 1984b: 228-229). Other disturbances may be more profound and thus result in the reorganization of the spatial relations in a field.

In amphibians the amputation of a limb usually results in its unmodified regeneration. Stock and Bryant (1981) have shown, however, that if amputation is performed at the level of separation of the digits, regeneration will result in a constrained variety of morphological patterns. Stock and Bryant (1981) devised a polar coordinate model to interpret their experimental results, assigning a symmetrical set of circumferential values to each developing digit which is dependent on the latter's relative position within the autopodium. While this model helps to interpret the constrained structure and potential variation of the vertebrate limb (at least in part: Holder, 1983), it renders the "exercise of defining detailed homologies among the digits of different species" rather "arbitrary" (Stock and Bryant, 1981: 431-432). Homology is here understood in its original topological sense.

A field, such as the limb bud, must be controlled by some organizing information programming the generative process. Form is created anew during each ontogeny according to this generative program; it is not the identity of constituent elements which is preserved throughout the cycle of reproduction, but the program according to which the elements form anew in every generation. The element of invariance is the generative program determining the morphogenetic field; it becomes expressed through the actualization of spatial relations and constraints between elements defining the morphogenetic field. Webster (1984: 206) has characterized the underyling generative mechanisms as "real essence" of an organism or of a group of organisms; the actualized pattern in contrast is the "nominal essence" involved in the reconstruction of taxonomies. The program of the field sets the limits of diversity within the constraints of a given spatial organization, and it is this constraint on potential diversity which renders similarity of form in terms of topological relations at all intelligible (Goodwin and Trainor, 1983: 91-92) and classification of organic diversity at all possible. The reverse of the medal is that potential diversity is constrained, i.e. that different potentialities of actualization are separated by impossible morphologies. For instance: the spatial organization of the limb-bud field permits the variable reduction of peripheral digits, but it will not permit a pattern in which the external digits involve more segments (phalanges) than the internal ones (Goodwin and Trainor, 1983: 92), although individual segments may be of larger size or greater structural complexity due to differential growth (Holder, 1983).

If transformation occurs, this must be due to some change in the program, which will define new spatial relations and constraints in the field, according to which the elements will be created during ontogeny. No genetic and positional identity of constituent elements is preserved during phylogeny; the elements will be created during ontogeny according to a new generative pattern, and this new pattern will be separated from the ancestral one by impossible morphologies. There will be gaps in the morphological continuum.

Goodwin and Trainor (1983) invoke the example of the ichthyosaur limb transformed into a paddle which is difficult to compare in detail with a normal tetrapod limb from an atomistic perspective. The proximal element and the two bones distal to it may be identifiable in spite of shortening, but problems of correspondence become acute beyond that point. Hyperphalangy and hyperdactyly must be invoked at least in typical forms (two poorly known genera approaching the primitive tetrapod condition have been described from the Lower Triassic of Japan and China: Carroll, 1987: 252-253) to account for the multiplication of constituent elements, and any identification of carpal or tarsal bones (as well as of the original digits) of the primitive pattern must be based on very shaky grounds if at all possible. Viewed from the generative paradigm, all attempts of such atomistic comparison are futile. Ichtyhosaurs are indeed a group of marine reptiles whose ancestry still remains shrouded in the past, and the absence of intermediates linking the group to "stem reptiles" was invoked as evidence for the inadequacy of the Darwinian research program by critics of Darwin already (Pictet, 1860). Viewed from a generative paradigm, gaps in the morphological continuum find their explanation by the modification of generative mechanisms resulting in the creation of a new morphology which cannot be compared to an ancestral one by the supposed genetic and positional identity of constituent elements. It is not the elements which are preserved and handed down through the genealogical nexus and which thereby permit the recognition of similarity in divergent appearances. Instead, it is the field program which constitutes the element of invariance underlying form, and modification of the "limb-generating process" may result in a great variety of limbs which are all transforms of the same basic pattern (Goodwin and Trainor, 1983: 76). Similarity, based on spatial relations of constituent elements, is a problem of form generation, not of descent.

The generative paradigm proposed by Goodwin (1984a,b) has two important consequences. Firstly, it would seem to alter the understanding of homology, and secondly it results in a holistic view of organization. Homology, as conceived by E. Geoffroy Saint-Hilaire, refers to constituent elements of structures compared in terms of positional relations. Homologous elements of two or more structures are believed to be reducible to an identical primordial element which appears differently following divergent metamorphosis. Implicit in this view is that identical primordial elements (constituting the type) are believed to underly the diversity of appearances. This is the implication which brought the concept of homology close to an atomistic background and permitted the explanation of the *unité du type* by the Darwinian model of atomistic transformation. In the perspective of the generative paradigm, constituent elements cannot be retained as such in transformations of a given structure; there is no material continuity of constituent elements throughout the process of transformation. Instead, they are created anew during each ontogeny, according to generative mechanisms controlled by a program for the developmental field. As a consequence, the comparison of organisms cannot be based on the identification of particular elements as being transformations of one another. Instead, homology must refer to the developmental program defining the spatial relations and constraints within the field. Just what the fields are, however, "is very much a research problem" (Goodwin, 1984b: 229).

By the same token, the generative paradigm is related to holism. The organism is not composed of separate parts, but develops epigenetically within a field or a combination of fields. Transformation does not affect parts of a given structure but structure as a whole: "biological patterns must be seen and understood as wholes which undergo transformation, so that there is no identity of elements in different patterns" (Goodwin and Trainor, 1983: 85).However, there obtains at this juncture a similar limitation of the model of developmental fields as it also characterizes the principle of connectivity advocated by Shubin and Alberch (1986; see chapter 3). To designate the limb bud as a developmental field presupposes the identification and hence the homology of the limb bud as a constituent element of the gnathostome type. This, however, can only be achieved on topological grounds. Whereas the explanation of the change from a pentadactyle limb to an ichthyosaur fin on the hypothesis of changing generative mechanisms of ontogenesis would account for the absence of intermediate stages of differentiation, the initial comparison of the different types of limbs presupposes their homology, and this presupposition must be based on topographic relations. In a similar sense, the claim that the spatial organization of the developmental field will prohibit the development of a pentadactyle limb with the peripheral digits longer than the medial ones presupposes topological criteria by the notions of "peripheral" as opposed to "medial". In the last analysis, the generative paradigm finds itself relegated to topographical criteria of similarity, too: without these, the delineation of a developmental field and the characterization of its spatial organization would be impossible.

The History of Form

In the context of atomism or Darwinism, form is treated as a functional problem. The unity of type becomes relevant in the context of its evolutionary explanation only. Diversity is a consequence of the process of descent with modification in adaptation to various environments: existing parts are modified (metamorphosed), new parts are added - processes which are continuously tested by natural selection. The creative power of natural selection is viewed as efficient cause of transformation: function constrains structure, evolution molds ontogeny. As the process of transformation runs through a series of small steps, permitting all possible intermediates between any two complex structures, the atomistic perspective is related to a serial arrangement or organisms implying a nominalist conception of types and groups.

Structuralism analyses form and its possible transformations independent of any hypotheses about process. Its research object are generative mechanisms, cast in terms of constraining morphogenetic mechanisms (Shubin and Alberch, 1986) or of a field theory of development (Goodwin, 1984a, b). Form has precedence over function, and the analysis of the mechanisms creating form during ontogeny has precedence over historical explanation. The structuralist 'way of seeing' relates to an essentialistic background, finding the essence of organisms, i.e. the developmental programs, translated into disjunct types and subtypes of organization. Essential similarity or homology ideally no longer relates to constituent elements and their topological relations within complex structures, but to the generative control mechanisms which define these relations. Ontogeny creates form, and therewith becomes the efficient cause of transformation. Selection may influence form within the boundary conditions defined by the constraining generative mechanisms, but it loses its powers to create new types of organization.

Whereas the Darwinian approach is related to a continuous serial arrangement of organisms, illustrating the genealogical relationships between ancestor and descendants, the generative approach emphasizes discontinuity as illustrated by the hierarchy of groups within groups resulting from the deviation or bifurcation of developmental programs or pathways. It will be recalled, however, that the two expressions of hierarchy are complementary to each other: a serial or linear hierarchy is translated into a subordinated hierarchy by the specification of inclusive taxa, whereas the subordinative hierarchy is changed to a serial arrangement of forms by the specification of actual ancestors (bifurcation of developmental pathways will introduce a branching point into the serial arrangement of forms). This complementary relation of hierarchies illustrates the fact that terminal addition to an existing ontogenetic sequence (of parts or of transformational steps) is but a special case of deviation (Lovtrup, 1978). The corollary of this chapter is that the functionalist versus the structuralist 'ways of seeing' are again complementary to each other. Analysis of structure, i.e. of patterns of similarity, will be related to the taxic approach, seeking the causes for structural diversity in the mechanisms of ontogenesis which determine a static hierarchy of groups within groups. Analysis of the process of transformation will be related to a functionalist point of view, asking how structures make life possible in a given environment, and what the possibilities are that environment may influence the transformation of structures in an adaptive sense. The structuralist approach focuses on the homogeneous primordium and follows its ontogenetic differentiation to heterogeneous complexity along deviating developmental pathways. The functionalist point of view starts from the environment and asks how the reproductive organism survives, and how it could improve survival. Where the two viewpoints meet, the compromise solution proposed by E. Geoffroy Saint-Hilaire and also emerging from the quotation of Alberch (1980) given above prevails: environment may create form, but only within the limits of the type as created by ontogeny.

CHAPTER 6

BEING AND BECOMING - THE CONFLICT OF PATTERN AND PROCESS

In the context of the "philosophical" or "transcendental" anatomy of E. Serres and E. Geoffroy Saint-Hilaire, the unity of type was meant to reveal law-like relations of form. In their view, lawful relations of similarity resulted from an active principle, some developmental force or *nisus formativus*, which would structure matter according to underlying principles. The most general principle in the analysis of form was the *principe des connexions*.

The intention did not change with changing explanations offered for the unity of design. Natural Theology evoked a metaphysical entity: the reduction of the diversity of organic appearances to a unity of type would lead back to the unity of the Creator as necessary and sufficient cause for the existence of law-like relations in the realm of nature. Evolutionism, on the other hand, explained the unity of type on the hypothesis of the unity of descent. Both traditions of thought are rooted in the desire to explain the multiplicity of organic appearances by some unifying principle serving as first and sufficient cause explaining the diversity of life on earth. Classifications as hierarchical systems of information storage and retrieval are an excellent example to show how a static and hierarchically organized system of notions, developed to master organic diversity, is used as basis for a causal theory about organic processes.

The comparison of these alternative traditions of thought might have provided the insight, however, that the search for unifying principles, unifying causes, and unifying explanations, is a futile endeavor. Looking at Creation myths it becomes obvious that they provide a perfect mirror of the structures of human cognition. To develop a notion of something, that thing must be placed in relation to something else in order to make comparison and therewith the relation of similarity or dissimilarity possible. A comparison as such presupposes opposites: there cannot be a notion of white without the opposing notion of black. This is why Creation myths introduce basic opposites into the world which is to serve as homestead of the most perfect of all Created beings, capable of rational cognition. The good is separated from and opposed to evil; darkness is separated from and opposed to brightness; land is separated from and opposed to the sea, the human being is opposed to animals, the woman becomes separated from and opposed to man.

The same basic operations can be observed in the process of classification. Goethe already, in his comments on the debate between Cuvier and Geoffroy Saint-Hilaire of 1830, clearly distinguished the two components of thought: Cuvier was characterized as working indefatigably towards a distinction and discrimination of organic beings by the recording of differences, while E. Geoffroy Saint-Hilaire was said to be seeking an underlying similarity of form in order to raise the knowledge of organic diversity to a higher level of cognition. Goethe's type concept was nothing but an attempt to combine these complementary intellectual operations. Distinction and discrimination among similar things is the operation which records taxic diversity - without taxic diversity, on the other hand, there is no need nor any reason to develop the concept of similarity; similarity in turn can give rise to taxic diversity by the process of transformation, so that without a canon of comparison taxic diversity would remain unexplained.

The same complementarity and interdependence of opposites persists into modern comparative biology. Continuity reveals itself to human cognition, i.e. it cannot be conceptualized other than as a series of discrete steps - however finely graded this may be. On the one hand, these discrete steps have to be related to each other on the hypothesis of continuity, while on the other hand the series can be translated into a complementary subordinated hierarchy of inclusive groups. The taxic approach records logical, i.e. static relations of homology specifying a hierarchy of groups within groups. The transfomational approach seeks to transform relations of coexistence into dynamic relations of succession, imposing a serial perspective on subordination. Pattern is opposed to process, yet the two are complementary 'ways of seeing'.

Pattern: The Reconstruction of Phylogeny

Phylogeny reconstruction by homology:

The recognition of a character, i.e. of a homology, minimally implies the comparison of two character-states or organisms: homology is a relational concept, and to recognize a homology means to recognize a group (Patterson, 1982). More precisely: the recognition of a homology boils down to a hypothesis of similarity, which in turn presupposes a conjectural group. Possible tests of hypotheses of similarity were discussed above (chapter 3). The relation of homology is linked to the hypothesis of hierarchy (Riedl, 1975; Brady 1983). The taxic approach establishes a subordinated hierarchy of groups within groups, which on the hypothesis of transformation becomes translated into a serial hierarchy specifying the ancestral and descendant conditions. In order to be empirical, the hypothesis of hierarchy must be testable, and this in turn raises the question of the focal level at which characters may be hypothesized to be homologous, i.e. to diagnose a group.

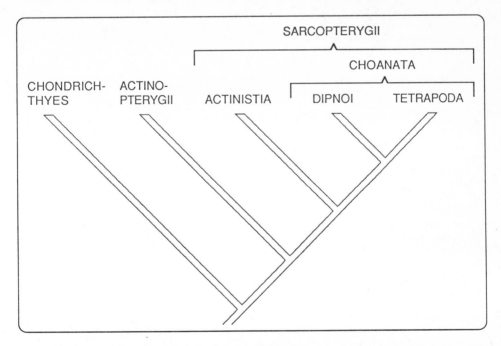

Fig. 11: The classification of the Gnathostomata (after Rosen, Forey, Gardiner and Patterson, 1981). For further discussion see text.

It was argued in the preceding chapter that taxic and transformational homologies stand in a subordinated relation to each other. The hypothesis of transformation must fall into a more general hypothesis of homology: the hypothesis that the tetrapod limb is to be derived from the sarcopterygian fin presupposes the homology of paired appendages of gnathostomes, and more specifically the sister-group relationship of sarcopterygian fishes and tetrapods. The "*Rahmenhomologie*" (Riedl, 1975) provides the *frame* for hypotheses of transformation. The tetrapod limb is a special homology, and therefore characterizes a group at a less inclusive level than the general homology of the paired appendages, which are a character of gnathostomes (see the synapomorphy scheme proposed by Rosen, Forey, Gardiner and Patterson, 1981). Paired appendages do not characterize a group composed of sarcopterygian fishes plus tetrapods, although both have paired appendages - but so have other gnathostomes. Jaws do not characterize a group composed of birds and mammals, although both have jaws - but so have other gnathostomes! Paired appendages and jaws form a group at a more inclusive level: they are special homologies of the Gnathostomata. At the subordinated level of the Sarcopterygii (including sarcopterygian fishes plus tetrapods), the 'paired appendages' have turned into a frame homology, while the sarcopterygian appendage is the special homology. The sarcopterygian appendage characterized as 'An exclusively metapterygial pectoral and pelvic fin...supported by a single basal element...' (Rosen, Forey, Gardiner and Patterson, 1981: 256) turns itself into a frame homology at the level of the Tetrapoda, because the tetrapod limb is a special expression of it: the sarcopterygian fin provides the frame for the hypothesis of transformational derivation of the tetrapod limb if tetrapods are recognized as the sister-group of the Sarcopterygii.

Homologies forming a group including *all* members characterized by the possession of the respective feature are special homologies of that group: the special homology constitutes the criterion of membership in that group (Rieppel, 1983a). In cladistic terminology special homologies are called synapomorphies, translated as "shared derived characters" (Hennig, 1967). By this translation, the term "synapomorphy" takes on an evolutionary connotation: as a shared derived character, it presupposes the identification of the primitive character or character-state. However, pattern analysis is concerned with logical relations of homology, and hence is static and ahistorical: the branching diagram (cladogram) indicates nothing but the generality of character distribution. This is why the term "synapomorphy" was synonymized with homology by Patterson (1982) - homology used at the correct level of inclusiveness and hence treated as "special homology" by Rieppel (1983a). When the term "synapomorphy" is used, the distinction of logical and hence ahistorical pattern analysis as opposed to the explanation of this pattern on the hypothesis of a phylogenetic process (Brady, 1985) must therefore be kept in mind.

Groups characterized by synapomorphies are called monophyletic by cladists. This terminology again implies a hypothesis of descent: the synapomorphy is interpreted as an evolutionary innovation, indicating a potential hypothetical ancestor, and the monophyletic group characterized by this evolutionary innovation is considered to include the ancestor plus *all* its descendants. Empiricists might want to restrict pattern analysis to the hierarchy of "natural groups" instead, i.e. to groups founded in nature and discovered by observation, whereby it is understood that observation must be guided by theoretical preconceptions and methodological rules, or tested on the basis of such rules (see chapter 3). In the given context, pattern analysis will be restricted to the discussion of *useful* groups. A useful group results from a hypothesis of grouping which strives for maximal information content. The name used to designate a useful group imparts maximum information on its members by the criteria of membership for that group (Rieppel, 1983a). Maximum information storage is achieved in a classification of "groups within groups" if all homologies are used as diagnostic features at the correct level of their inclusiveness. Useful groups are thus based on maximal congruence of characters.

Groups including some, but not all organisms characterized by some (homologous) character are based on a frame homology (Riedl, 1975). In other words: the respective homology is used at the wrong level of inclusiveness. The group is useless because it does not include all members characterized by the respective homology which constitutes the criterion for membership (Rieppel, 1983a) in that group. Or conversely, to characterize a group including sarcopterygian fishes plus tetrapods by paired appendages is senseless, because paired appendages constitute the criterion of membership in a more inclusive group: at the level of sarcopterygian fishes and tetrapods, the criterion of membership is wrong.

Frame homologies are called symplesiomorphies in cladistic theory, a term which - like synapomorphy - carries a phylogenetic connotation. Plesiomorphy designates the phylogenetically primitive condition from which the apomorphous condition of the character in question is thought to be derived. In cladistic terminology, groups based on a frame homology are dubbed paraphyletic. This implies that the frame homology characterizes a potential hypothetical ancestor (the ancestor of the Gnathostomata in the above example), and that the paraphyletic group - based on this character used at the wrong hierarchical level - includes the ancestor but not all its descendants (it includes the sarcopterygian fishes plus tetrapods, but not actinopterygian fishes, acanthodians, elasmobranchs and placoderms which all share paired appendages).

It is easy enough, however, to free pattern analysis from the phylogenetic connotations implicit in the terms symplesiomorphy and paraphyly. As should have become clear from the simple example given above, each special homology turns into a frame homology at less inclusive levels of the hierarchy, or, to put it in other words: a frame homology is nothing but a special homology related to the wrong level of inclusiveness in pattern analysis! Once again: 'paired appendages' is a frame homology (symplesiomorphy) at the level of sarcopterygian fishes and tetrapods, but it turns into a special homology (synapomorphy) at the level of the Ganthostomata. Jaws are a special homology (synapomorphy) at the level of the Gnathostomata and thus constitute a criterion of membership in that group. In phylogenetic perspective, the Gnathostomata include the ancestor - "inventor" of the evolutionary innovation of jaws - plus all its descendants. Used as a criterion of membership in a group comprising birds plus mammals, the jaws are a frame homology (symplesiomorphy), and the group - if based on this character - is paraphyletic: it must include the "inventor" of jaws (since jaws are present in mammals and birds), but it does not include all its descendants.

Since every frame homology turns into a special homology if used at the correct higher level of inclusiveness (since every synapomorphy turns into a symplesiomorphy if used at a lower level of inclusiveness), it is possible for pattern analysis, i.e. for the taxic approach, to dispense with the distinction of frame and special homology, of symplesiomorphy and synapomorphy, and thus to get rid of evolutionary connotations. Each symplesiomorphy becomes a synapomorphy and hence a special homology if used at the correct level of inclusiveness, and it is at this level only that homologies form membership criteria of useful groups. The distinction of frame and special homology, of symplesiomorphy and synapomorphy, becomes obsolete in the taxic approach, and synapomorphy turns out to be a synonym of (taxic) homology (Patterson, 1982). We are talking of nothing but homologies or characters forming groups, and the task of the taxic approach is reduced to finding the correct level of inclusiveness at which a given homology or character forms a *useful* group. It is only in the context of the transformational approach that the distinction of frame and special homology becomes necessary. It is only in the context of process analysis, i.e. if pattern is to be explained on the hypothesis of descent, that the polarization of characters in a phylogenetic sense, as expressed by the terms synapomorphy and symplesiomorphy becomes reasonable and necessary. Indeed, the hypothesis of transformation creates this distinction in the first place. The character 'An exclusively metapterygial pectoral and pelvic fin…supported by a single basal element…' (Rosen, Forey, Gardiner and Patterson, 1981: 256) forms a group including the sarcopterygian fishes plus tetrapods. If the tetrapod limb is accepted as a special differentiation of the sarcopterygian fin, explained as being phylogenetically derived from the latter, the character 'An exclusively metapterygial pectoral and pelvic fin...supported by a single basal element...' (Rosen, Forey, Gardiner and Patterson, 1981: 256) turns into a frame homology, plesiomorphous at the level of the Tetrapoda. It is the translation of a taxic (subordinated) hierarchy into a transformational series of forms which creates polarization in a phylogenetic sense by the fact that terminal taxa on the branching diagram, belonging to the same inclusive taxon (and hence shown to be related to each other), are forced into a temporal (phylogenetic) sequence. In short: as transformational homologies are subordinated to taxic homologies, every hypothesis of transformation presupposes a hypothesis of relationship.

Theory of cladistics distinguishes not only monophyletic and paraphyletic groups, but also polyphyletic ones. The commonly accepted definition of polyphyletic groups is that of J. Farris (as quoted by Wiley, 1981: 84; see also Schoch, 1986: 181): "A group in which the most recent common ancestor is assigned to some other group and not to the group itself". The concept of polyphyly is again cast in terms of process, and is as such an interpretation of some hypothesis of grouping based on pattern analysis. The polyphyletic group is recognized, i.e. hypothesized, on the basis of membership criteria which have been wrongly assigned the status of homologies. A polyphyletic group is based on convergent characters, which is why it fails the test of phylogeny (of congruence).

Pattern reconstruction critically depends on the level at which similarity can be hypothesized as a homologous relation diagnosing some useful group. If this is not achieved, the threat of paraphyly or polyphyly hampers the phylogenetic explanation of pattern, giving rise to faulty assumptions about transformational processes. It is therefore necessary to have methods at hand which will allow the assignment of homologies to the correct focal level of analysis in pattern reconstruction. Several such methods have been proposed, and they will be discussed in sequence. A major problem is, however, that most of these methods have not been formulated independently of assumptions about process, as is already indicated by their designation as methods of character polarization. This creates circularity if the explanatory hypothesis of phylogeny is to be evaluated (tested: Brady, 1985) against the pattern of order in nature. It will therefore be necessary to attempt to free the various methods of character polarization from phylogenetic connotations.

The first method to be discussed is the criterion of character correlation. In its original version the method postulates that a character in a group under consideration is likely to be primitive if it occurs in correlation with other primitive characters in that group; of course, the polarity of the other characters would have to be determined independently. Expressed as above, the method is entirely embedded in a process perspective. It is falsified by character incongruence. Explained on the hypothesis of phylogeny, the conflicting distribution of characters must indicate mosaic evolution: characters evolve independently of each other, so that any organism may be expected to share a mixture of primitive and derived characters. The classic example is *Archaeopteryx*, the alleged ancestor of higher birds (deBeer, 1954). *Archaeopteryx* does not document a harmonious and gradual transition from the reptilian grade of organization to the type of higher birds; rather, it exhibits a mixture of primitive reptilian features with avian characteristics such as the presence of feathers and of a furcula (fused clavicle). From a pattern point of view, the so-called primitive features characterize a group at some more inclusive levels, while feathers, furcula and other characters are special homologies (synapomorphies) uniting *Archaepopteryx* and birds in a group of higher rank, thereby indicating their relatedness.

In the present context another example will be chosen, namely the tuatara or *Sphenodon*, usually considered as a living fossil and a close relative of modern lizards. The similarity between the tuatara and modern lizards is so striking that the first specimen of *Sphenodon* to be described was referred to the agamids by J.E. Gray in 1831. Only in

1840 did R. Owen recognize the similarity of *Sphenodon* to Mesozoic fossils, while in 1867 A. Günther made the point that the tuatara had wrongly been classified with modern lizards (see Robb, 1977, for a more detailed historical account). Today, *Sphenodon* is considered the only living representative of the Sphenodontida, a group including Mesozoic fossils and considered to be the sister-group of the Squamata, comprising lizards, amphisbaenians and snakes. Both spenodontids and squamates belong to a more inclusive taxon, the Diapsida, characterized by two fenestrae in the temporal region of the skull. The upper temporal arcade, made up from the postorbital and squamosal bones, separates the two temporal fenestrae from each other, while the lower temporal arcade, made up from jugal and quadratojugal bones, provides the ventral closure of the lower temporal fossa.

As compared to modern lizards, *Sphenodon* is characterized by a number of primitive skeletal and soft tissue characteristics, amongst which the complete lower temporal arcade figures prominently; the latter has been reduced in modern lizards, where the quadratojugal bone is completely missing (to the possible exception of some geckos: Brock, 1932). The reduction of the lower temporal arcade in modern lizards permitted movement of the jaw suspension (streptostylic movement of the quadrate bone) in correlation with a complex mechanism of cranial kinesis (Frazzetta, 1962). In *Sphenodon*, the quadrate bone is immovable, firmly held in place by the lower temporal arcade.

The tuatara also shows a type of tooth implantation known as acrodont (the tooth is attached to the edge of the supporting bone), which amongst modern lizards is only observed in agamids and chamaeleons (some amphisbaenians are also acrodont). In correlation with other primitive features, the acrodont tooth implantation was interpreted as yet another primitive feature of *Sphenodon*, so that it came as a surprise when the complex mode of tooth growth, replacement and function was described in detail (Robinson, 1976). Tooth wear predicted a type of jaw movement which did not appear primitive either, although quite different from the type of movement observed in modern lizards. These predictions were confirmed, and further details of the jaw mechanics described, in a recent study by Gorniak, Rosenberg and Gans (1982), who showed *Sphenodon* to perform complex translational movements of the lower jaw in a powerful shearing bite. Among extant reptiles the tuatara is the only genus for which translational movements of the lower jaw have been demonstrated by experiment. The animal is known to be a successful predator: it is able to kill seabirds and, in one instance at least, a rat has been found as stomach content (Robb, 1977).

The tricky thing about these studies on the dentition and the function of the jaws is that they give rise to the supposition that the complete lower temporal arcade of *Sphenodon* may not be primitive, but redeveloped in order to brace the quadrate bone, immobilize the jaw suspension and thus make the powerful shearing bite possible. This is in accordance with the observation that in the generalized diapsid condition, the jugal and quadratojugal bones meet below the lower temporal opening, whereas in *Sphenodon* the jugal forms an extended posterior process, providing the entire lower margin of the inferior temporal fossa, meeting the quadratojugal behind the latter; the quadratojugal bears no anterior process. Furthermore, the embryos of the tuatara show rudiments of cranial joints (Howes and Swinnerton, 1901), an observation which might point to the secondary immobilization of the skull. These speculations find further support in recent developments in paleontology. Indeed, new fossil material of early diapsids, including fossil sphenodontids, indicate that the Sphenodontida were primitively characterized by an incomplete lower temporal arcade, and that on the basis of cladistic analysis those sphenodontids which have a complete lower temporal arcade must have redeveloped it secondarily (for details and summary of the literature see Carroll, 1985, and Whiteside, 1986). The polarization of characters by correlation has thus been falsified by pattern analysis.

Another method of character polarization, again based on an evolutionary perspective, is that of ingroup comparison: "A character state that is restricted to the group of organisms being studied is primitive when it is usually or frequently exhibited by the individuals of that group. The probability that a character state is primitive increases very rapidly with the increase in the number of individuals in the group that exhibits that state" (Kluge, 1971: 26). This principle, also known as commonality principle and stating that the character-state which is more common in a group under investigation is primitive as compared to the less commonly observed condition, has been discussed and refuted by Watrous and Wheeler (1981). An elegant example illustrating the inadequacy of the commonality principle or ingroup comparison is the status of the Monotremata as compared to other mammals (Schoch, 1986). The Monotremata are known from two genera and four species which are primitive in many features as compared to marsupials and placentals. Monotremes lay eggs, yet they have hair and suckle their offspring. In some osteological features, they resemble reptiles more closely than mammals: they have a sprawling gait, and the shoulder girdle retains such primitive elements as an interclavicle and separate precoracoids and coracoids. Judged by ingroup comparsion these features must be considered derived as they are less frequently observed within mammals. The fact is, however, that the features in question do not form a group within the Mammalia, i.e. the Monotremata, but instead are special homologies (synapomorphies) at a more inclusive level, characterizing a more inclusive group including reptiles (synapsids and even more) and mammals. It is the absence of these features in all other mammals (viviparous, erect gait, no interclavicle) which characterizes a subgroup within the Mammalia, i.e. the Eutheria.

A character cannot be shown to be plesiomorphous simply by its correlation with other, primitive, character-states. Nor can it be judged plesiomorphous on the principle of commonality. To frame the problem in assumption-free pattern analysis terminology: the character which within a group under investigation is less commonly distributed, does not necessarily specify a subgroup of the taxon under study as compared to more widely distributed characters within that group. The monotreme example given above is particularly suited to illustrate this point. However: how do we know that the presence of an interclavicle, although rare among mammals, is a homology at a more inclusive level than that of the Mammalia? This knowledge can only come from outgroup comparison, i.e. from a comparison with reptiles which generally retain the interclavicle.

Outgroup comparison is indeed a logically stringent method for the determination of "character polarity", i.e. of the level of inclusiveness at which a homology characterizes a group. Its logical stringency results from the fact that it corresponds to the method of downward classification by logical subdivision, as already advocated, albeit in a crude version, by Georges Cuvier (Rieppel, 1987b). Cuvier's method of pattern analysis proceeded from the diagnosis of more inclusive groups to that of less inclusive, subordinated groups by successive dichotomization. An example for the application of that method is provided by Duméril's (1806) classification of reptiles.

At that time, the amphibians were not separated from the class Reptilia. Nor was Duméril able to free his mind entirely from the concept of the *échelle des êtres*. He therefore numbered the terminal taxa on his branching diagrams so as to indicate their relative position in a serial or linear classification (Rieppel, 1987b). The two alternative classifications of reptiles which he proposed in his *Zoologie Analytique* were based on two different sets of characters.

Duméril (1806) characterizes the Reptilia by the presence of lungs and ectothermy, but deprived of hair, feathers and mammary glands. Within the Reptilia, he finds the possibility to introduce a first dichotomy on the basis of presence (character $\underline{a}$ in Fig. 12) or absence ($\underline{a}$*) of limbs: this groups the Ophidia as opposed to all other reptiles. These in turn are dichotomized on the basis of presence ($\underline{b}$) or absence ($\underline{b}$*) of claws, a character which separates lizards and turtles from batrachians. Lizards and turtles finally become separated on the basis of presence ($\underline{c}$) or absence ($\underline{c}$*) of teeth.

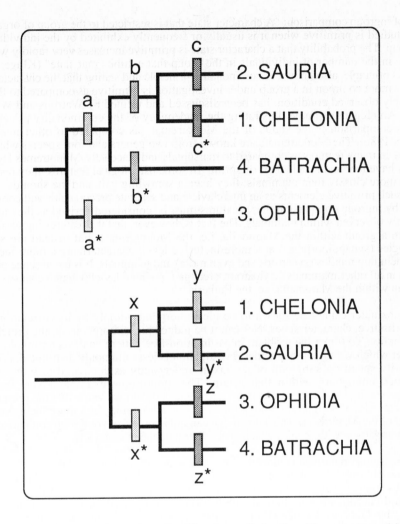

Fig. 12: Two alternative classifications of the Reptilia as proposed by Duméril (1806), based on Cuvier's method of successive dichotomization. The numbering of the taxa corresponds to their placement in a serial arrangement. For further explanations see text.

An alternative basal dichotomy may be introduced into the Reptilia on the basis of heart structure: Cuvier, and his collaborators Brongniart and Duméril, believed the heart of turtles and lizards to have paired auricles (character x in Fig. 12), while that of snakes and batrachians was thought to be characterized by the presence of a single auricle only (x^*). Turtles have a carapace but no teeth (y), whereas lizards have teeth but no carapace (y^*). Ophidia have a squamous skin but no limbs (z), whereas batrachians are naked but have two or four legs (z^*).

Consider the argument rather than its defectuous results as judged by modern standards. For example, the first solution separates the lizards from turtles by the presence of teeth. Applying the principle of commonality, one would have to conclude that teeth are more widely distributed within reptiles, and hence cannot serve to characterize a subgroup within the latter. Their presence can, however, be used to diagnose the lizards as opposed to turtles if reference is made to an explicit hypothesis of grouping at a more inclusive level, putting both lizards and turtles into a common group of higher rank. A similar argument applies to the alternative hypothesis: the characterization of batachians *inter alia* by the presence of limbs would not seem to pan out in view of the fact that limbs are more widely distributed in reptiles. Yet their presence does diagnose batrachians if the character is related to a higher level hypothesis putting snakes with batrachains in the same group.

One point should be made at this juncture: in Duméril's classifications of reptiles (1806) many of the groups (by Duméril's own standards all those characterized by an asterisk in Fig. 12) are based on the absence of characters. Aristotle designated such groups as "privative", and on this basis criticized Platon's method of dichotomization, because privative groups - based on the absence of characters - cannot be further subdivided (Gigon, 1982). Privative groups are also rejected in modern cladistic theory (Eldredge and Cracraft, 1980; Nelson and Platnick, 1981), because privative groups are paraphyletic. Amphibians, if not characterized by a special homology of their own (autapomorphy), are nothing but Tetrapods without amnion. As such, the Amphibia are characterized by homologies which form a group at a more inclusive level, i.e. the Tetrapoda. In his classification of the Reptilia, Duméril starts out with exactly such a privative group: his reptiles are characterized either by the absence of characters, or by features such as lungs and ectothermy which form groups at a more inclusive level!

Privation is different from secondary loss, however, in a process orientated point of view: if snakes are recognized as tetrapods on the basis of *other* features that relate them to lizards, it becomes possible to *explain* the absence of limbs by reduction. A privative group is characterized by the simple absence of a feature; secondary loss implies that something was added before it can be reduced. In Duméril's classification of reptiles, not all groups characterized by the absence of characters are to be considered privative by modern standards (*sensu* Eldredge and Cracraft, 1980, and Nelson and Platnick, 1981), because the absence of teeth (characterizing turtles) and the absence of limbs (characterizing snakes) can be explained, from a process-oriented point of view, as a result of reduction rather than of privation. It is the lizards which are paraphyletic, because teeth form a group at a more inclusive level, and it is the group comprising the Sauria plus Chelonia plus Batrachia which again is characterized by a homology used at the wrong focal level, since limbs diagnose the Tetrapoda.

The point which the discussion of Duméril's (1806) classification of reptiles was intended to bring out is that according to Cuvier's method, dichotomization proceeds downwards, i.e. from more inclusive to less inclusive levels of the subordinated hierarchy, each taxon being explicity diagnosed with reference to a hypothesis of grouping at the next higher rank. If, on the basis of other characteristics, lizards and turtles are recognized as members of a more inclusive group, then the presence versus the absence of teeth can be used as a diagnostic feature within that group. Downward classification is also the characteristic of the outgroup method, which therefore is comparable to Cuvier's method for its logical stringency, although privative groups are no longer admitted by modern theory (they are frequently admitted in modern practice, however). Outgroup comparison is used to determine the level of inclusiveness at which characters form unambiguous groups. The analysis proceeds in a downward direction, starting from the most inclusive levels and specifying successively less inclusive groups by successive dichotomization. As a group of higher rank ($\underline{A}$) becomes subdivided or dichotomized ($\underline{a}$, $\underline{b}$), the resulting subgroups stand in a sister-group relation to each other. If dichotomization continues with ($\underline{b}$), producing ($\underline{b}$, $\underline{b}$*), a hierarchical subordination of groups within groups obtains: (A) contains a minimal hierarchy of ($\underline{a}(\underline{b},\underline{b}*)$): ($\underline{a}$) is the sister-group and therewith the outgroup of ($\underline{b},\underline{b}*$); ($\underline{b}$) in turn is the sister-group of ($\underline{b}$*). The sister-group relations specify that ($\underline{b}$) is closer to ($\underline{b}$*) than either is to ($\underline{a}$).

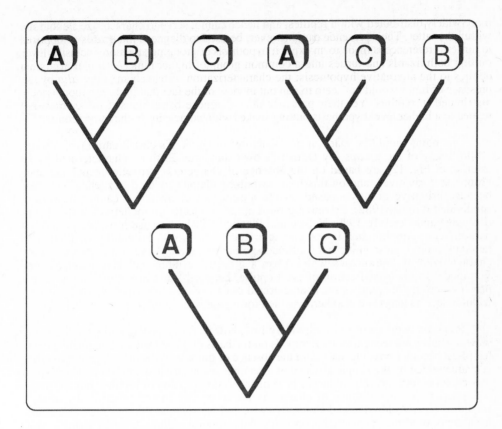

Fig. 13: A three-taxon statement and its three possible permutations.

Homology denotes the logical relation of form, and the principle of outgroup comparison is the method used to reconstruct the hierarchy of homologies on a logical basis. Consider the basic statement implicit in any subordinated classification, not only in biological ones: "A is closer to B than either is to C" (or any permutation thereof). The basic statement therefore involves three taxa (A, B and C), and it specifies the relation of these three taxa by stating that two are closer to each other than either is to the third. The basic classification thus corresponds to a "three-taxon statement" (Lovtrup, 1977).

Every three-taxon statement can generate three possible permutations, which permits easy resolution in case of character incongruence (see chapter 3). It is appropriate, therefore, to assess the level of inclusiveness at which a given character or homology forms a group within a three-taxon statement. That most classifications will comprise more than three terminal taxa is no hindrance to this procedure, since every classification, as long as it is fully dichotomous, can be resolved into a nested hierarchy of three-taxon statements, each involving three basic taxa (Gaffney, 1979). The cladogram of gnathostomes presented by Rosen, Forey, Gardiner and Patterson (1981: Fig. 621; see Fig. 11 above) may serve as a case in point. The Gnathostomata comprise two groups, one representing the sister of the other: Chondrichthyes and Osteichthyes. A three-taxon statement results at the next lower level of inclusiveness, at which the Actinopterygii pair off from the Sarcopterygii within the Osteichthyes. The basic classification involves the taxa Chondrichthyes, Actinopterygii and Sarcopterygii. The Sarcopterygii divide into the Actinistia and Choanata at the next lower level, so that a three-taxon statement obtains involving Actinopterygii, Actinistia and Choanata. The Choanata comprise the Dipnoi and Tetrapoda, so that the basic classification at this level involves the Actinistia, Dipnoi and Tetrapoda. The procedure could be continued into the Tetrapoda, which subdivide into Amphibia and Amniota, the latter breaking down into Sauropsida and Mammalia, etc.

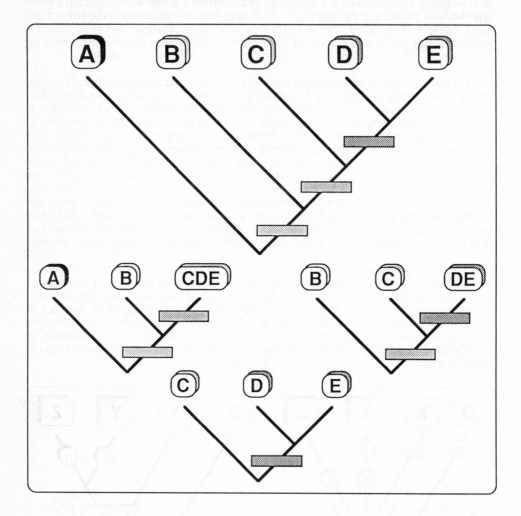

Fig. 14: The decomposition of a subordinated (fully dichotomous) classification into a series of internested three-taxon statements. For further discussion see text.

Note that as pattern reconstruction proceeds down through the subordinated levels of the hierarchy, the outgroup relations become successively specified. The Chondrichthyes provide the outgroup for a comparison within Actinopterygii and Sarcopterygii; the

Actinopterygii provide the outgroup for a comparison of Actinistia with Choanata; the Actinistia are the outgroup for the analysis of interrelationships of dipnoans and tetrapods; dipnoans finally are the outgroup for the analysis of tetrapod interrelationships.

The analysis of relations within a three-taxon statement thus serves to establish the level of inclusiveness at which a homology characterizes a group with respect to a given hypothesis of grouping at the next higher rank, specifying the outgroup to the group under study. Some authors admit several outgroups into pattern analysis, but even then it is specified that the comparison with the sister-group of the taxon under investigation is the "most critical" (Wiley, 1981: 7), and it alone will provide unambiguous results.

Consider the following hypothetical example. There are three taxa (X, Y and Z) which belong into a group specified by homology B. The three taxa stand in an unspecified relation to each other and constitute the ingroup under analysis. With reference to gnathostomes, the question to answer might be whether actinistians (X) or dipnoans (Y) are the sister-group of tetrapods (Z). That the three taxa belong into a group of higher rank would be specified by Sarcopterygian homologies or synapomorphies (homology B in Fig. 15).Two characters (a and b) are observed which both pass the test of similarity (chapter 3). On this information alone, the three taxa continue to partake in an unresolved trichotomy. Neither character generates a hypothesis of grouping which would resolve the trichotomy, nor can the principle of commonality be applied, not only because of theoretical insufficiency, but also because both characters are of equal frequency within the group under investigation. The trichotomy cannot be resolved because on this information alone there is no way to decide at which level the characters in question (a and b) form a group.

To get on with pattern reconstruction, the outgroup (O) must be introduced into the analysis. This is done with reference to a hypothesis of grouping at a more inclusive level, specifying that O is the sister-group of the taxon (X,Y,Z), the two - and only the two - belonging to a more inclusive group characterized by homology A. With reference to gnathostomes, it would be the Actinopterygii which would constitute the sister-group and hence the outgroup of the ingroup (Sarcopterygii) under investigation, and the character (A) specifying this sister-group relationship would be an osteichthyan homology or synapomorphy.

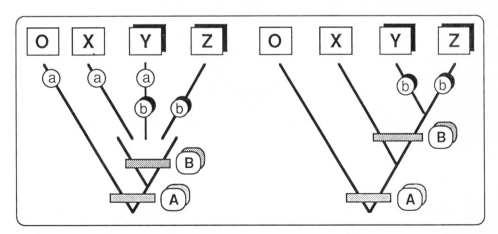

Fig. 15: The method of outgroup comparison. See text for further discussion.

With the outgroup being specified we are now in a position to analyze character distribution with respect to a more inclusive level of the hierarchy. Character a is shown to occur both in the ingroup and in the outgroup, whereas character b does not. This demonstrates that character a really forms a group at a more inclusive level of the hierarchy, and therefore cannot be used to specify a taxon subordinated to the group (X,Y,Z). Were it used to this end, it would generate a paraphyletic group. Character b on the other hand is missing in the outgroup. We can therefore assume that at the focal level of analysis, character b does not specify a hypothesis of grouping at a more inclusive level, but rather represents a homology, characterizing a taxon subordinated to the taxon (x,y,z), i.e. to the Sarcopterygii.

Returning to the gnathostome classification proposed by Rosen, Forey, Gardiner and Patterson (1981), character 'a' might indicate the presence of lepidotrichia. This is a feature diagnosing the Osteichthyes, but not the tetrapods. Character 'b' might indicate the presence of a choana: it is present in lungfish and tetrapods, but not in actinistians nor in the Actinopterygii. Looking at Sarcopterygians alone, lepidotrichia would group Actinistia and (fossil) Dipnoi together, whereas lungfish and tetrapods pair off by the shared presence of the choana. On the basis of this information alone there is no way to choose between the alternative hypotheses of grouping.

As the sister-group of the Sarcopterygii - the Actinopterygii - becomes specified, it is immediately clear that lepidotrichia characterize a group at a more inclusive level, i.e. the Osteichthyes, comprising both, the Actinopterygii and the Sarcopterygii. The choana on the other hand is absent in the Actinopterygii and thus properly characterizes a subgroup of the Sarcopterygii, the Choanata.

The discussion of outgroup comparison also makes clear why the principle of commonality would not work (Watrous and Wheeler, 1981). At a first glance this principle might seem to convey a correct observation, i.e. that the more commonly distributed character specifies a more inclusive group inserted further down on the axis of the cladogram; the character in question therefore is to be interpreted as frame homology or plesiomorphy from a transformist's point of view. The principle fails because it does not relate commonality of occurrence to a specified segment of the hierarchy under investigation. It corresponds to ingroup comparison, illustrated with the marsupial example given above.

The situation can be remedied with the specification that a character is considered to be more common if it ocurrs in the ingroup under investigation *plus* in its sister-group, i.e. in the outgroup (Rieppel, 1984a). By that specification, the principle of commonality turns into what was defined as 'principle of generality' by deQueiroz (1985: 283; following a personal communication by W.P. Maddison): "character x is more general than character y if and only if all organisms possessing x (at some stage in ontogeny) also possess y and in addition some organisms possessing x do not possess y". The level of generality is the level at which a homology characterizes a group - a monophyletic group if viewed from a historical perspective - and it is determined by means of outgroup comparsion.

This, however, is the point which has evoked so much criticism against outgroup comparison. Resolution of interrelationships within a group under study depends on the assessment of the level of generality at which a character diagnoses a group, and this in turn depends on the acceptance of a hypothesis of grouping at a more inclusive level, specifying the sister-group and therewith the outgroup of the ingroup under investigation. Higher level relations must be hypothesized before lower level relations can be worked out properly. This is downward classification, and it proceeds by logical subdivision on a deductive base of argumentation. From some hypothesis of higher level grouping the level of generality of character distribution used to reconstruct lower level subdivisions is deduced. The critical point is that this type of deductive reasoning necessitates the *a priori* acceptance of a more inclusive hypothesis of grouping (Wiley, 1981; Rieppel, 1983a; Schoch, 1986: 138), an apparent deficiency of the method highlighted by a number of critics, evolutionists (Mayr, 1982) and empiricists (Patterson, 1983: 21-23) alike.

However, this cannot be considered a serious weakness of the method of outgroup comparison. First of all it must be pointed out once more that there cannot be any assumption-free method or theory in the realm of natural sciences: even a simple and reproducible experiment makes a number of *a priori* assumptions with respect to boundary conditions and instrumentation (Feyerabend, 1981). If that is so at the level of singular statements, how could any general statement such as a scientific theory be assumption-free (Popper, 1976a)? In fact, the dependence of every test on some assumptions lying beyond the theory under test was one of the important criticisms of early versions of falsificationism (Chalmers, 1986). The point, therefore, cannot be whether or not *a priori* assumptions should be acceptable; instead it must be required that these assumptions are made explicit, and are themselves open to test or evaluation. These requirements are met by the method of outgroup comparison. Each conjectured outgroup relation is not only, by its specification, made explicit in a hierarchical system of groups within groups, but it is itself open to evaluation (test) with reference to the next higher level in the hierarchy and by the introduction of new characters into the analysis.

Authors rejecting outgroup comparison because of its dependence on *a priori* assumptions of relationship must propose alternative ways of pattern reconstruction which would promise to avoid this (unavoidable) pitfall. The method of upward classification proposed by evolutionists (Mayr, 1982) is certainly not convincing, as it has recourse to *a priori* assumptions about the process of transformation which are not open to independent evaluation (test: Brady, 1985). Character weighting (discussed in chapter 3) is but one case in point! Patterson (1982, 1983) on the other hand proposes ontogeny as the only empirical (observational) key to the level of generality at which a character forms an unambiguous group. This is an issue to be dealt with in the next paragraph, where it will be shown that recourse to ontogeny in pattern reconstruction again is not assumption-free. It is just different assumptions which are being accepted.

Kluge and Strauss (1985: 252) found the method of outgroup comparison to be "explicitly evolutionary because it requires a hypothesis of phylogeny (particular outgroups, plus the ingroup)...", and they used this perception as an argument against the independence of systematics. The above discussion should have made clear, however, that outgroup comparison is not dependent on evolutionary theory, but only on more inclusive hypotheses of grouping. The evolutionary explanation of these hypotheses of grouping is a different problem!

The dependence on higher level hypotheses of grouping cannot justify the rejection of the method of outgroup comparison - but why should it be accepted? The method provides a logically stringent tool for the representation of conjectured homologies in a subordinated hierarchical pattern. It is a way to properly assess the level of inclusiveness at which a conjectured homology forms a useful group. However, not all conjectured homologies fall into the same subordinated pattern of 'groups within groups'. Asking for the most complete hierarchy of homologies within a given set of topological relations of similarity implies the problem of conflicting character distribution. Not all topographical correspondences fall into the same hierarchical pattern; several alternative hypotheses of grouping can be established, from which the most parsimonious one must be selected (see chapter 3). The most parsimonious classification is the most *useful* one.

The question "Why Classify Organisms" was answered by Kluge (1971: 37) in a particularly perceptive way:

"A classification provides the necessary framework for the mental storage of vast amounts of information into a condensed and readily retrievable form. The nomenclature associated with the classification also provides a standardized terminology that can be used to communicate objectively with others about the information embodied in the classification. In addition to these practical aspects, a classification may in itself provide the stimulus for further conceptualization..."

If we ask for the best hierarchy contained within the data set, the parsimony principle follows necessarily (Brady, 1983). Within a given set of topological relations of similarity containing some degree of incongruence, the parsimony principle will maximize congruence, i.e. it will minimize the occurrences, on the cladogram, of those characters which have passed the test of similarity. The most parsimonious cladogram corresponds to the simplest representation of the hierarchical distributions of topographical correspondences and therefore maximizes information storage with respect to the properties of the taxa included. Maximal information content will thus be conveyed to notions introduced to designate groups (Nelson and Platnick, 1984). We are not asking for "real" groups, but for the most useful groups instead. Useful groups must be unequivocal, i.e. they must be diagnosed by homologies used at the correct level of generality; and the most useful groups are those which have the greatest information content.

The "usefulness" of a hypothesis of grouping may not only be measured by its information content, but also by the explanatory power of the genealogical hypothesis related to it. The hypothesis of grouping explains nothing in itself. However, the genealogical hypothesis based on a most parsimonious classification can be considered to represent the one which is best supported by observation, which therefore is the most likely one: it is the genealogical hypothesis with the greatest explanatory power, because it explains the greatest number of characters by inheritance from a common ancestor (J.S. Farris, quoted and discussed in Sober, 1983).

It is only by the addition of the time dimension and of the notion of ancestry that homologies can be interpreted as evolutionary novelties characterizing natural groups, while incongruent characters are interpreted as instances of convergence. The intention for generating such an explanation is the search for the sufficient reason of the *unité dans la variété*. Whether phylogeny is not only a sufficient, but also a necessary (Brady, 1983) explanation for the hierarchy of homology is another problem (Rieppel, 1983a). Let Russell (1916: 302-303) conclude this paragraph:

"We have seen that the coming of evolution made comparatively little difference to pure morphology, that no new criteria of homology were introduced ... The principle of connections still remained the guiding thread of morphological work ... the natural system of classification was passively inherited, and, by a *petitio principii*, taken to represent the true course of evolution".

Phylogeny reconstruction by ontogeny:

Pattern reconstruction must be based on the concept of homology; the problem, as discussed in the preceding paragraph, is to determine the level of generality at which a homology diagnoses an unequivocal group. In cladistic terminology this problem is called the determination of character polarity, a terminology which for most authors has an evolutionary connotation but does not necessarily carry it (deQueiroz, 1985: 285). What is being asked for is nothing but the level of generality at which a homology forms a useful group. "It seems, therefore, that ontogeny is the decisive criterion in determining polarity...", writes Patterson (1982: 54), specifying: "Phylogeny is generalised transformation, but we have no empirical experience of phylogeny; the only transformations of which we have empirical evidence are those of ontogeny" (Patterson, 1983: 21). These statements are complex and require detailed analysis step by step.

First of all, Patterson's is an empiricist's claim: phylogeny cannot be observed! It is the reconstruction of the hierarchy of homologies which provides the empirical basis for phylogenetic hypotheses. Ontogeny, on the other hand, can be observed, and if ontogeny would provide a clue to the hierarchy of homologies, it might be interpreted so as to open a window into the past. What has happened during phylogeny would in principle become observable and thus amenable to empirical investigation during ontogeny.

Correlated with this call for empiricism is a rejection of outgroup comparison because it presupposes hypotheses of grouping at a more inclusive level than the focal level under analysis (see above). "What criterion is to be used to establish the ingroup and the outgroup?", asks Patterson (1983: 22), and continues: "In other words, without some prior knowledge of character phylogeny, how can we investigate phylogeny?" The point is that outgroup comparison is considered to be theory laden, whereas the ontogeny method for phylogeny reconstruction is considered to be empirical, based on observation with no *a priori* assumptions about phylogeny. Nelson (1985) has called the outgroup comparison an "indirect method", as it presupposes a hypothesis of relationship, whereas ontogeny would provide a "direct method" for the reconstruction of the genealogical hierarchy.

The discussion of outgroup comparison above has made clear, so it is hoped, that it does not imply assumptions about process, but assumptions about the level of generality at which characters form groups. Outgroup comparison presupposes the more inclusive group, comprising the outgroup plus the ingroup under investigation, in order to permit the latter's subdivision: the method is one of downward classification by logical dichotomization. In contrast, the ontogenetic method is offered as an empirical guide to the correct level of inclusiveness at which a character forms a group, and hence as an assumption-free guide to common ancestry. Two separate parts of the argument must therefore be distinguished. The first concerns the reconstruction of the hierarchy of homologies on the basis of outgroup comparison or ontogeny. The second argument concerns the explanation of pattern by the process of phylogeny. The questions to be asked at this juncture are: a) whether the ontogenetic method is indeed assumption free, and b) which side of the argument is affected by such assumptions as are eventually implied.

That the ontogenetic method should be assumption-free has recently been denied by Kluge (1985: 14; see also Kluge and Strauss, 1985): "That there exists some positive relationship between similarity and homology seems to be an unavoidable assumption in these procedures". Assumptions come in at a very basic level of argumentation, which only demonstrates that it is futile to search for any assumption-free statements. Alberch (1985: 47) has further emphasized that the use of the ontogenetic method in pattern reconstruction is based on a "particular conceptualization of ontogeny which does not always correspond with the truly dynamic nature of development". Conceptualization is involved at two levels: the concept of homology requires that a continuous developmental process be subdivided into a sequence of discrete and therefore discontinuous stages (Rieppel, 1985c), and these stages in turn must be comparable among ontogenies of different organisms as well as between ontogenies and adult stages. One is reminded of Charles Bonnet's concept of the "*Tout organique*", based on the principle of continuity. If development is a continuous process of differentiation, with the preceding stage constraining the next one and at the same time containing within it the boundary conditions set by the next stage to develop from it, how is any stage to be delineated from any other for the purpose of comparison, unless on a largely arbitrary basis (Kluge and Strauss, 1985: 253)? And if this process of ontogenetic differentiation is one of continuing individualization, the more general giving rise to the less general and hence to the more particular, how can an ontogenetic stage of one individual organism be compared with the adult stage of another (von Baer, 1828)? The use of ontogeny as a direct method of pattern reconstruction presupposes the concept of homology and with it the "*principe des connexions*" of Etienne Geoffroy Saint-Hilaire. Constituent elements of organic structures must be abstracted from the developing "*Tout organique*" if the dynamic process of ontogenetic differentiation is to be translated into a static hierarchy of types (Rieppel, 1985c). So much for the argument about classification!

If, beyond mere pattern reconstruction, it is assumed that the process of ontogenetic differentiation mirrors the process of phylogenetic development, this second part of the argument critically depends on the conservation of ancestral characters in descendant ontogenies (deQueiroz, 1985; see also Kluge, 1985, on the possibility of dedifferentiation). A distinction has been made by Alberch (1985) of causal versus temporal ontogenetic sequences (see chapter 3), a distinction roughly corresponding to Lovtrup's (1978) concept of epigenetic vesus non-epigenetic characters. Causal ontogenetic sequences are preserved by developmental constraints such as inductive interactions: the development of the gut roof must precede the development of the neural tube, because the latter is induced by the first (deBeer, 1958: 149-150). (Note that the establishment of the homology of the neural tube on this inductive mechanism presupposes the homology of the gut roof.) In a causal ontogenetic series, "stage $\underline{A}$ is required for the expression of $\underline{B}$, and, in turn, $\underline{B}$ is a prerequisite for the expression of $\underline{C}$" (Alberch, 1985: 49). In contrast, a temporal ontogenetic sequence is unconstrained, the "lack of causality between the antecedent and the descendant stages" rendering the sequence useless for systematics (Alberch, 1985: 49). It is the characters of a temporal ontogenetic sequence which are prone to incongruent distribution, with the effect that the ontogenetic method will work for causal ontogenetic sequences only (Lovtrup, 1978; Alberch, 1985). However, an assumption is being introduced at this stage of argumentation, namely that the causal chain of events generating characters during ontogeny will not change, i.e. remain the same between organisms. Yet, such changes have been demonstrated (Alberch, 1985: 49, for the induction of the lens in *Rana esculenta* as opposed to *Rana fusca*; see also chapter 3).

The discussion of pattern and process as revealed by ontogeny again necessitates the distinction between taxic and transformational relations of homology in the analysis of ontogenetic stages or sequences. It is now generally accepted that the ontogenetic method is originally linked to von Baerian recapitulation (Patterson, 1982, 1983), of which Haeckelian recapitulation is but a special case (Lovtrup, 1978). Von Baer's laws of individual development emphasized two basic points (see chapter 4): a) ontogeny does not recapitulate adult stages of lower forms (laws three and four of von Baer, 1828), and b) ontogeny is a continuous process of differentiation, producing heterogeneity from homogeneity, whereby it is observed that the characters of more inclusive groups of animals precede the more restricted characters of subordinated groups in development (law one), and that the more universal condition of form gives rise to the more particular (law two). It is true that von Baer was an Aristotelian essentialist (Kluge, 1985; see also chapter 4) dealing with the abstract and static notion of types, but that did not hinder him from finding the hierarchy of the types paralleled in the laws of individual development: he viewed each organism as an individual, but also as a member of a type and subtype. No process of transformation of one type into another is implied (Rieppel, 1985c), but the hierarchy of types is mirrored in the hierarchical organization of the "evolutionary" (ontogenetic) process of differentiation.

Von Baer's first law predicts that the characters of types develop in sequence along the lower edge (axis) of the cladogram (see chapter 4). This is falsified by non-terminal insertions of larval or embryonic adaptations (Kluge, 1985): the amnion develops prior to the features characterizing the Tetrapoda (Patterson, 1983: 25). This was also known to von Baer (1828: 204) who used the amnion as one amongst many arguments against the Meckel-Serres Law of terminal addition (see chapter 4). The first law holds, however, in cases of conservative causal ontogenetic sequences, involving structures (epigenetic characters *sensu* Lovtrup, 1978) characterizing the type in its fully differentiated condition (in the "*bleibende Thierform*", as von Baer, 1828: 206, would have it). The first law is a special case of the second law, resulting from the causal interdependence of developmental sequences in the whole organism.

The second law is a guide to sequence in structural relations (Patterson, 1983): the more universal gives rise to the more particular. As an Aristotelian, von Baer (1828) defended an epigenetic conceptualization of the ontogenetic process. What his second law specifies is that structures necessarily develop from and out of one another, the first to be formed providing the material cause for the next one to develop. Beyond these structural relations the law specifies that the process of ontogenetic differentiation proceeds from more generally distributed features to less generally distributed structures. The original formulation of von Baer's second law is couched in a rather abstract language: homogeneity precedes heterogeneity. The most simple and universal condition of form is the vesicle (von Baer, 1828: 258); maximum complexity is developed in man. This abstract terminology permits two alternative readings of the law, i.e. reading ontogenetic sequences in taxic or in transformative terms. "No jaws precedes jaws", "no amnion precedes amnion", "no visceral clefts precedes viceral clefts, which precedes viseral clefts closed" (transformation involved), "no hyomandibular or stapes precedes presence of either" (Patterson, 1982: 54): this reading implies taxic relations of homology, specifying the origin of features characterizing groups what adds to heterogeneity in the hierarchy of types within types. The differentiation of jaws renders the type of the Ganthostomata more heterogeneous than the type of the Vertebrata, and so does the differentiation of limbs which is preceded by "no limbs". This type of reading has been criticised on the ground that 'nothing' cannot precede 'something' in the epigenetic process of differentiation. In fact, however, the statement that "no amnion precedes amnion" is but an abbreviated version of Patterson's (1983: 25) statement: "The embryonic membranes of amniotes, as developments (outgrowths) or [of?] more widely distributed structures..."

No doubt some primordial tissue must be present to give rise to the amnion or to jaws, as is lawfully required by the epigenetic mode of development. It is this more widely distributed primordial tissue which is shared by a more inclusive group, which therefore is more universal; the amnion is a particular differentiation thereof, characterizing a less inclusive group.

So far, the second law of von Baer has been discussed from the perspective of the taxic approach. There also exists the possibility to read it in a transformational sense, although that would violate its original meaning. Instead of stating "no jaws precedes jaws" or "a more universally distributed embryonic primordium gives rise to jaws in some, but not all, organisms which share it", one might state that a mandibular arch transforms into jaws in some, but not all, organisms that have the mandibular arch. This interpretation of the second law involves a transformational homology, the mandibular arch of agnathans giving rise to the jaws of gnathostomes. If the homology of the mandibular arch with jaws is not accepted, e.g. because of the different position of the structures in relation to the gill filaments (Moy-Thomas and Miles, 1971: 5), there is no basis for the claim that the mandibular arch is more universally distributed among vertebrates than are the jaws. A necessary condition for reading von Baer in a transformative sense is the homology of embryonic structures in one organism with adult structures in another organism. Beyond this point, the conjecture of a transformational homology presupposes a hypothesis of grouping on which to base the claim for phylogenetic relationship - quite apart from the theoretical problems involved in the comparison of adult structures with embryonic rudiments. The claim that the mandibular branchial arch transformed into jaws during the phylogeny of the Vertebrata presupposes the inclusion of the Agnatha and Gnathostomata within a common taxon of higher rank, and it implies the homology of the embryonic rudiments of gnathostome jaws with the mandibular arch of adult agnathans. This, however, is plainly a misinterpretation of von Baer's (1828) "Laws of Individual

Development", which were meant to characterize a process of individuation. What von Baer would have stated is that agnathans and gnathostomes share a common embryonic *anlage* of the splanchnocranium, characteristic of the type Vertebrata. Deviation sets in as Agnathans, as during their perfection they develop a mandibular arch from that primordium which in gnathostomes develops into jaws. The primordium, in this case, would be the neural crest cells from which both, the mandibular arch of agnathans (Koltzoff, 1902, quoted in deBeer, 1937: 473) and the jaws of gnathostomes (Hall, 1978: 39) develop. Deviation or divergence of their migratory pathways with respect to the visceral clefts would explain the different position of adult structures with respect to the gill filaments.

Patterson (1983: 25-26) finds the second law of von Baer, read in terms of the taxic approach, not falsified by any observation in vertebrate morphology. deQueiroz (1985: 283) specifies, however: "...if the hierarchy of groups is determined by the sequence of character transformations in ontogeny [alone], then this law is an unfalsifiable tautology. Von Baer's law is testable only if the hierarchy of groups and, thus, the generality of characters is determined according to some other criterion". Furthermore, ontogenetic transformations of primordial tissue into adult structures carry the problem that they are frequently not observable: Kluge (1985: 14) discusses the origin of the primary appendicular musles in amniotes in comparison to other vertebrates as an example. In vertebrates with the exception of amniotes, appendicular musculature develops from the lateral trunk musculature, i.e. form a defined set of myotomes. In amniotes, the muscles originate from an undifferentiated blastema in the limb-bud. But even if transformations are observed, these observations frequently represent generalizations from the investigation of separate stages (deQueiroz, 1985: 297). This only legitimates the more clear-cut and radical formulation: "no jaws precedes jaws", "no limbs precedes limbs", "no amnion precedes amnion". No transformation is implied by these statements; they only specify a hierarchy of groups within groups. Indeed, much criticism of the use of the ontogenetic method in phylogenetics confounds the two sides of the argument: pattern analysis preceding the explanation of pattern by process.

The charge of tautology raises the question of which "other criterion" might be used to put Patterson's (1983) reading of the second law of von Baer to test? One such independant criterion would seem to be provided by outgroup comparison, but then the question must be answered how knowledge of the appropriate outgroup is acquired (Nelson, 1978; Patterson, 1982): it is the ontogenetic method which is claimed to sort out ingroups and outgroups in the first place! Consider an example popular with 18th century biology. The crystal as much as the "zoophyte" *Hydra viridis* were believed to form, i.e. to develop from a combination of originally equivalent "primary morphic units". But the development (or ontogeny) of the polyp differs from the development (or ontogeny) of the crystal in that it results in *irritability*, a trait absent in the latter formation. Ontogeny, proceeding from absence to presence, thus helps to resolve ingroups and outgroups by telling us that living things can be separated from inorganic matter, and in so doing it provides a start for a subordinated classification. A subordinated hierarchy of types (Rieppel, 1985c) results from the resolution of ingroups and outgroups by the ontogenetic method.

The problem is to decide whether the second law of von Baer is both necessary and sufficient to determine the level of generality at which a character forms an unambiguous group. A problem of interpretation of the original wording is involved here, but it certainly can be stated that von Baer did not differentiate between the principle of commonality and that of generality. If pattern is reconstructed on the basis of the commonness of character states in ontogenetic sequences, the analysis is in fact based on ingroup comparison (Kluge, 1985: 24) and suffers from inherent inadequacies. This is the reason why Nelson (1978: 327) restated the biogenetic law (von Baer's second law) as follows : "given an ontogenetic character transformation, from a character observed to be more general to a character observed to be less general, the more general character is primitive and the less general advanced".

By this formulation, von Baer's second law becomes related to the principle of generality (de Queiroz, 1985) and thus should in principle yield a hierarchy of (taxic) homologies which is congruent with the genealogical hierarchy based on outgroup comparison. If the absence of a character represents the more generalized "condition of form", while the differentiation of a structure results in a less general "condition of form" (to use von Baer's verbiage), it must follow that the ontogenetic method does indeed provide a direct access to a resolution of ingroups and outgroups. This claim has been criticized by a number of authors (Rieppel, 1979a; Kluge, 1985; see also Fink, 1982, de Queiroz, 1985, and Alberch, 1985, for further discussion) on grounds of parsimony, a criticism which again confounds pattern and process, however.

The debate will here be illustrated with an argument first developed by Lundberg (1973), rebutted by Nelson (1973) and taken up again by Kluge (1985), followed by a renewed defense by Nelson (1985).

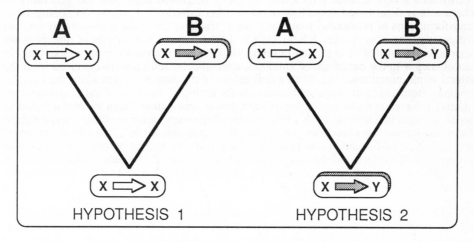

Fig. 16: The relation of parsimony to ontogeny. For explanation see text. (After Kluge, 1985: Fig. 6).

Let there be given two taxa (Fig. 16), taxon A showing character x, taxon B possessing the characters x and y. Ontogenetic differentiation proceeds from x to y. For two taxa there exists only one possible cladogram, but on it there are different possible representations of the characters in question (Nelson, 1985). The first hypothesis would interpret character x as a gain in the hypothetical common ancestor of the taxa A and B, while character y becomes differentiated in taxon B only. Taken by itself the ontogenetic method would indicate that character y diagnoses a taxon at a less inclusive level: every organism which shares character y also shares character x, but not all organisms sharing character x also share character y; taxon B is subordinated to the taxon [AB]). This first hypothesis implies two steps: the gain of character x in the hypothetical common ancestor of the taxon [AB], and the gain of y in taxon B.

A second hypothesis would assume the characters x and y to have been present in the ontogeny of the hypothetical common ancestor of the taxa A and B. On that assumption, character y must have been secondarily lost in taxon A, for instance by paedomorphosis. This hypothesis implies three steps: the gain of the characters x and y in the hypothetical common ancestor of the taxon [AB], and the subsequent loss of character y in taxon A. The first hypothesis is more parsimonious, the ontogenetic method therefore proves successful (Nelson, 1985). This is correct, but, as Kluge (1985) notes, such a decision seems to depend on 'prior knowledge' in that it presupposes knowledge of the sister-group of the taxon [AB] indicating the presence of character y in the hypothetical common ancestor of that taxon. The ontogenetic method therefore does not seem to be independent of outgroup comparison. Ontogeny must be interpreted, according to Kluge (1985; see also Rieppel, 1979a), in the light of the hierarchy of homologies as it obtains from outgroup comparison. Only such comparison would indicate whether character y represents a gain in taxon $\underline{B}$, or whether it must have been lost by paedomorphosis in taxon $\underline{A}$.

It is true that ontogeny by itself, as the only indicator of hierarchy, cannot detect paedomorphosis (Rieppel, 1979a; Fink, 1982; Kluge, 1985). However, this criticism does not affect the use of the second law of von Baer in pattern analysis, because heterochrony is a statement about phylogeny (Gould, 1977), not about classification! A simple example, derived from the ontogeny and classification of snakes (Rieppel, 1979a), may serve to illustrate the issue. Snakes have been recognized as squamate reptiles ever since the work of Oppel (1811). sister-group of the Squamata is the tuatara, *Sphenodon punctatus*, and its fossil allies; the Sphenodontida and the Squamata together form the Lepidosauria. This hierarchy of groups within groups is supported by a large suite of characters, but there is incongruence in one important feature, viz. the development of the anterior base of the braincase. In vertebrates generally, the base of the anterior (prechordal) division of the neurocranium is laid down as paired condensations, the trabeculae cranii. In all extant reptiles, but also in fossil "stem reptiles" (as indicated by a narrow anterior process on the parasphenoid in captorhinomorphs and *Procolophon*: Heaton, 1979; Carroll and Lindsay, 1985) the trabeculae cranii fuse to form a trabecula communis, resulting in the tropibasic skull (character y in Fig. 16). Snakes are the only exception: they retain paired trabeculae and hence a platybasic skull (character x in Fig. 16).

On the basis of that character alone as judged by the ontogenetic method, snakes would have to be classified outside the Lepidosauria (a conclusion arrived at by Hoffstetter, 1955, 1962, 1968, and Rage, 1982), or even outside the Reptilia. It is only if their classification as squamate reptiles is accepted on grounds of parsimony that the platybasic condition of the snake skull can be recognized as an instance of character reversal due to paedomorphosis. The detection of paedomorphosis is thus dependent on a hierarchical classification established on grounds of parsimony. But this classification need not necessarily be constructed by outgroup comparison. Classification is, in the first instance, a question of order versus disorder, of character congruence versus incongruence (Patterson, 1982), and the test of character congruence can be applied not only to outgroup comparison, but also to the hierarchy of ontogenetic differentiation. By the absence of limbs, the snakes might seem to pair off with eels; on the basis of their platybasic skull they seem to fall outside the Reptilia. But if ontogenies are compared with respect to many features, one or a few being paedomorphic, most having remained unaffected by heterochrony, the totality of the characters under investigation would provide a parsimonious solution (Fink, 1982: 262). If the ontogenies of *Sphenodon*, of a lizard and a snake were compared with each other, the comparison involving a multitude of characters, the snakes would turn out as squamate reptiles on grounds of congruence, i.e. of parsimony. The platybasic skull would represent an instance of character incongruence (with respect to higher levels of classification). The crucial point to note is that the ontogenetic method, just as much as the method of outgroup comparison, is prone to character incongruence. The parsimony principle must serve as Ockham's razor whichever method is applied in pattern reconstruction. The ontogenetic method is not tautological. The test of the ontogenetic method is the same as that for relations of homology established by outgroup comparison: the congruence of characters. So much for classification.

The causal explanation of character incongruence is another argument. Incongruence resulting from outgroup comparison is explained by convergent evolution; incongruence revealed by the ontogenetic method is explained by heterochrony (Gould, 1977). Paedomorphosis as implied in the example given above can have three causes (Gould, 1977, Alberch *et al.*, 1979): the first is progenesis, produced by precocious sexual maturation, so that the organism attains sexual maturity at a juvenile stage. The second type of paedomorphosis is neoteny, produced by retardation of somatic development so that the adult organism retains juvenile traits. The third instance of paedomorphosis results from post-displacement, related to a delayed onset of embryonic development. To distinguish between the three types of paedomorphosis, a rigorous hypothesis of sister-group relationships of the taxa concerned is required.

In summary, then, it can be concluded that classification is a problem of logical subordination. It can be arrived at by the indirect method of outgroup comparison, or by the direct ontogenetic method. As neither method can avoid character incongruence, the parsimony principle must in both cases serve to maximize congruence and thus help to search for the most complete hierarchy contained within the data set at hand (Brady, 1983). Heterochrony is a historically contingent statement about phylogeny, providing a hypothetical causal explanation for character incongruence as revealed by the ontogenetic method. The second law of von Baer provides a direct clue to the hierarchy of classification, the relation of taxic homology being subject to the test of congruence. Because of the possibility of heterochronic development, ontogeny cannot provide a direct clue to the phylogenetic process, however. Instead, the phylogenetic information content of ontogenetic sequences of differentiation must be deduced in the light of a hypothesis of hierarchical classification. Notions such as "prior gain" (in a hypothetical common ancestor) and "subsequent loss" (in some hypothetical descendant) are valid in their proper context only, i.e. as hypothetical causal explanation of observed character incongruence.

Process: Evolution as an Explanatory Hypothesis

"It is generally acknowledged that all organic beings have been formed on two great laws - Unity of Type, and the Conditions of Existence. By unity of type is meant that fundamental agreement in structure, which we see in organic beings of the same class, and which is quite *independent* of their habits of life. On my theory, unity of type is *explained* by unity of descent. The expression of conditions of existence, so often insisted on by the illustrious Cuvier, is fully embraced by the principle of natural selection. For natural selection acts by either now adapting the varying parts of each being to its organic and inorganic conditions of life; or by having adapted them during long-past periods of time..." (Darwin, 1859 [1959: 176-177]; italics added).

The above quotation, taken from Charles Darwin's book on evolution, is particularly revealing as it once again documents his views of the relations between systematics and evolution. Here, we have the law of the unity of type, which is independent of the conditions of existence and hence from the principle of natural selection: pattern analysis therefore is *independent* of evolutionary theory (Brady, 1985). The unity of type becomes *explained* by the unity of descent, that is: pattern becomes explained by the theory of evolution. Evaluating unity of type against conditions of existence, Darwin believed to be able to state that "...in fact, the law of the Conditions of Existence is the higher law; as it includes, through the inheritance of former adaptations, that of Unity of Type" (Darwin, 1859 [1959: 177]). Brady (1985) pointed out that if anything, it is the relation of homology which is empirical, whereas the hypothesis of descent ("inheritance of former adaptations") is not observed but added to pattern in order to explain it. Russell (1926: 246-247) summed up the situation as follows:

"The aim of ... pre-evolutionary morphology had been to discover and work out in detail the unity of plan underlying the diversity of forms, to disentangle the constant in animal form and to distinguish from it the accessory and adaptive. The main principle upon which this work was based was the principle of connections, so clearly stated by Geoffroy...The current morphology, Darwin found, could be taken over, lock, stock and barrel, to the evolutionary camp".

Darwin carefully distinguished the relation of homology from adaptation. The concept of homology is based on the principle of connectivity, which abstracts from form and function; in contrast, the law of "Conditions of Existence" is concerned with precisely these aspects of structure, viz. shape and function in terms of adaptation to biotic and abiotic environmental conditions. Opponents of the hypothesis of transformation of species were eager show that the unity of type, or the pattern of order in nature, did not necessarily follow from the law of "Conditions of Existence" alone. Louis Agassiz for instance, in his *Essay on Classification* (1859), was quick to point out that the law of "Conditions of Existence" would not explain why a microcosm such as a drop of water, surely a homogeneous medium, should contain animal life of great diversity: "The smallest sheet of fresh water, every point upon the sea-shore, every acre of dry land teems with a variety of animals and plants" (Agassiz, 1859: 15). Conversely the concept of adaptation left unexplained why the same type should preserve its essential structure and similarity in representatives which occupy different ecological niches or show widely disjunct or worldwide distribution. "As much as the diversity of animals and plants living under identical physical conditions shows the independence of organized beings of the medium in which they dwell, so far as their origin is concerned, so independent do they appear again of the same influences when we consider the fact that identical types occur everywhere upon earth under the most diversified circumstances" (Agassiz, 1859: 21). It is undeniable that parts of animal organization can be influenced by environmental conditions: "...the blind fish, the blind crawfish, and the blind insects of the Mammoth Cave in Kentucky, furnish uncontrovertible evidence of the immediate influence of those exceptional conditions upon the organs of vision" (Agassiz, 1859: 19); but why is it that the relations of these animals are not completely obscured by such environmentally induced modifications of organization. Why is it that rudimentary organs of no known function (with respect to the environment) do not completely disappear? "These and similar organs are preserved in obedience to a certain uniformity of structure, true to the original formula of that division of animal life, even when not essential to its mode of existence" (Agassiz, 1859: 12).

The quintessence is: "...in all these animals and plants there is one side of their organization which has an immediate reference to the elements in which they live, and another which has no such connection; and yet it is precisely that part of the structure of animals and plants, which has no direct bearing upon the conditions in which they are placed in nature, which constitutes their essential, their typical character" (Agassiz, 1859: 47). This is the aspect of organization seized upon by modern structuralism, investigating the causes of the distinctiveness of the type and explaining it by developmental constraints. The concept of an idealistic *bauplan* is thereby replaced by the empirical concept of generative mechanisms of form (Shubin and Alberch, 1986).

The essential similarities, characterizing the type and its subtypes, are rooted not in contingent environmental adaptations, but in the relation of homology. Homology, based on the principle of connections, is "a logical relation, independent of any historical or genealogical relationships which the actual structures may have" (Goodwin, 1984a: 101), giving rise to a rational taxonomy (Goodwin, 1982: 51). The cladogram, which graphically represents the hierarchy of homologies as it obtains from downward classification by logical subdivision, is likewise "atemporal" (Patterson, 1983; see also

Platnick, 1977). It becomes temporalized only by the addition of hypotheses of ancestor-descendant relationships, i.e. by the addition of the hypothesis of descent with modification. Because the hierarchy of homologies is atemporal by its logical nature, Agassiz (1859: 4-10) could raise the question whether order in nature might be nothing but a "pure invention of the human mind". He answered the question by the postulate of a preestablished harmony between the subordinated structure of classification and the structure of conceptual thinking (Agassiz, 1859: footnote to p. 9): "The human mind is in tune with nature, and much that appears as a result of the working of intelligence is only the natural expression of that preestablished harmony" (Rieppel, 1988b). The hypothesis of a preestablished harmony holds only with reference to some superior principle which would guarantee the isomorphy of the structure of conceptual thinking with the subordinated structure of classification. This principle, according to Agassiz, was God, and the existence of order in nature was evidence of a rational plan of Creation. God had Created - and still continues to Create - the world according to a rational scheme of order, while at the same time he endowed man with powers of cognition which made it possible to perceive order in nature and thus find proof for the existence of God. According to Agassiz, God's thinking was coextensive with nature which, together with the hypothesis of a preestablished harmony relating nature to human thinking, resulted in the important corollary that naturalists are able to correctly recognize "natural groups" whether or not they believe in God (Agassiz, 1859: footnote to p. 8).

As divine thinking is coextensive with nature, the reality of "natural groups" is founded both in the material world as well as in the ideal plan of Creation. Agassiz emphasized how "from a careful investigation of the subject for several years past" (Agassiz, 1859: 244) he was able to base his classification on an empirical basis, since "It is a fact so universal, in every sphere of intellectual activity, that practice anticipates theory..." (Agassiz, 1859: 207). On the other hand, he admitted that reality was imparted on species and supraspecific taxa by the fact of their Creation, since natural groups existed in the Creator's mind "before the first individual produced by sexual connection was born" (Agassiz, 1859: 253). Thus there resulted a curious duality in Agassiz's conception of reality: species "exist in nature in the same manner as any other group; they are quite as ideal in their mode of existence as genera, families, etc., or quite as real..." (Agassiz, 1859: 256).

In Agassiz's view, natural groups are both, an ideal category of divine thought as well as a physical entity in the hierarchy of nature. This corresponds to the conception of the hierarchy of homologies: as a logical relation and as a historical contingency of the material world. For Agassiz, the hierarchy of homologies existed as a timeless category of thought in the divine plan of Creation, a Plan which became successively materialized during the succession of geological epochs in the course of earth history (Agassiz, 1841: 13, 16; 1845: 7; 1859: 221-226, 256-257). The species represents a type on a low level of the hierarchy of classification, Created according to a logical scheme of relations: as such, it belongs to the 'world of being'. It is represented - not constituted! (Agassiz, 1859: 256) - by individuals which are borne, grow, become sexually mature and "are bound to some limited home during their lifetime" (Agassiz, 1859: 257) - individuals which, in other words, belong to the 'world of becoming'. The static type cannot be constituted by individual organisms which are themselves continuously changing. But the relations of topological similarity are observed in these individual organisms, which therefore represent the type. Throughout the cycle of reproduction, the essential characteristics of the type are preserved, because "evolution" (meaning ontogeny in this context) was restricted to limits "within appointed cycles of growth, which revolve forever upon themselves, returning at appointed intervals to the same starting-point and repeating through a succession of phases the same course" (Agassiz, 1874 [1973: 431]). The coextensiveness of an ideal category of thought with ever becoming entities of nature obviously necessitated the permanence of the type during the cycle of reproduction.

Agassiz recognized homology as a logical relation of similarity and hence as an element of 'being' in the 'becoming' of organisms. It is on the basis of its logical structure that the hierarchy of homologies can be claimed to represent "the only law-like hierarchic relationship in biology above the level of the individual organism" (Patterson, 1982: 22), specifying a hierarchy of 'groups within groups'. Agassiz explained it on the hypothesis of a rational plan of Creation, modern evolutionism explains it on the hypothesis of descent with modification.

Congruence has been called the "criterion of success" (Brady, 1985: 119) for conjectured homologies and their explanation on the hypothesis of descent with modification. However, the test of congruence relates only to the causal explanation of topological relations of similarity as evidence of common ancestry. In the face of conflicting distribution of characters which have passed the test of similarity, parsimony must guide the choice among several alternative hypotheses of grouping, favoring the most useful one, i.e. the one with the highest information content. Following the principle of parsimony, it is the maximally congruent characters which are the most trustworthy indicators of common ancestry, while incongruent features are attributed to convergent evolution.

Dullemeijer (1980: 176-177) has criticized pattern analysis for its adherence to the parsimony principle, which renders the cladogram an idealistic scheme: the reason for this criticism is that phylogeny does not necessarily follow the most parsimonious path. However, nobody has ever argued such a thing: parsimony follows from the search for the best hierarchical fit of the data (homologies), and in that context carries no biological implications (Brady, 1983). Dullemeijer's criticism adds something to the parsimony principle used in pattern analysis, namely the hypothesis of phylogeny, the concept thus becoming one of "evolutionary parsimony" (Dullemeijer, 1980: 176). It is only in the context of the evolutionary interpretation of a cladogram that the parsimony principle may cause problems (Rieppel, 1987c).

The cladogram is idealistic not by the use of parsimony, but by the very nature of the homologies on which it is based, as Dullemeijer himself in fact recognized, specifying: "Simple idealistic homology, which is what …synapomorphies in essence are, cannot be used to show phylogenetic relationship without *adding* [sic!] information about real blood relationship, genetic connection and time direction and time spacing…" (Dullemeijer, 1980: 177; emphasis added) . However, it is precisely this information which is added, if homology is explained on the hypothesis of descent. Dullemeijer (1980: 177) recognized the logical structure of the dichotomous cladogram: "Human thinking operates best in binary systems (see computer work) and any phenomenon can be described is such a binary system", but he added: "we must leave out Darwinian evolution and not include terms like ancestry". Do we? In some way, the cladogram and the pattern analysis on which it is based are indeed incompatible with Darwinian evolution and with the notion of ancestry. The cladogram does not specify actual ancestors, and by its hierarchical structure, emphasizing discontinuity of discrete classes based on similarity versus dissimilarity, it is incompatible with a gradual process of Darwinian evolution. Continuity implies all possible intermediates between any two structures; the hierarchy of homologies implies discontinuity between sharply demarcated types.

It may be appropriate at this juncture to recall that the study of evolutionary mechanisms is restricted to localized populations and therewith is related to the ecological hierarchy, whereas the hypothesis of descent with modification is based on an extrapolation, on the principle of uniformity, of the action of evolutionary causes into the phylogenetic past (see chapter 5). Whereas the study of mechanisms requires continuity in the cycle of reproduction, the extrapolation of causes of evolutionary change into the phylogenetic past presupposes the reconstruction of taxic relations of similarity. If nature does not appear to be ordered, there is no need for any evolutionary theory, because there is nothing to be explained (Brady, 1985). The conflict of the two contrasting viewpoints originates with the fact that pattern analysis, i.e. the taxic approach, emphasizes discontinuity and hence the discreteness of "individual" levels of the genealogical hierarchy, whereas the study of mechanisms of evolutionary change implies continuity and the notion of actual ancestors.

The Notion of Ancestry

The comparison of the Aristotelian *scala naturae* with the von Baerian concept of a branching order of nature has made clear that only the former arrangement of organic appearances is compatible with the doctrine of a gradual, i.e. continuous transformation of one organized being into another. This is because privative groups can easily be conceptualized as actual ancestors: the relation of descent involves nothing more but the (terminal) addition of new qualities (characters) to an ancestral ontogenetic sequence. The notion of ancestry does not permit the individuation of ancestral taxa, since this involves the differentiation of unique characteristics diagnostic of the respective group. The differentiation of such unique features (autapomorphies) would have to be reversed if the corresponding taxon is to serve as an ancestor. Ancestry from an individualized taxon must be explained by ontogenetic deviation, i.e. by non-terminal addition of new features to a stage of development which remains privative with respect to the descendant condition. From the fact that the relation of descent does not permit the individuation of an ancestral taxon results the problem that an ancestral group of organisms cannot as such be objectified unequivocally. Complete pattern analysis, serving as a guide to common ancestry, requires the unequivocal specification of inclusive taxa (by synapomorphies) as much as of exclusive taxa (by autapomorphies). Only at this stage have paraphyletic groups been completely eliminated. This, however, means that no ancestor can be objectified as a monophyletic group. The notion of monophyly is incompatible with the notion of ancestry (Nelson and Platnick, 1984).

Viewed from an evolutionary perspective, an actual ancestor (Haeckelian ancestral group: see chapter 4) must be primitive relative to its descendant in all its characters, i.e. it must be a paraphyletic taxon. If the ancestor is allowed to share some character of its own, diagnosing it as a group (at species level or below), reversal must be assumed for that character if it is to be absent in the descendant. If the ancestral character is also present in the descendant, it no longer characterizes the ancestor as an unequivocal group, but forms a group at a more inclusive level instead - including the ancestor and its descendants. On the other hand, if reversal is admitted, anything can in principle become ancestral to anything else, and the notion of ancestry degenerates about a statement on likelihood of reversal. Arguments about actual ancestry may take on the nature of "just-so stories" (Forey, 1982) or they may have recourse to the notion of "evolutionary parsimony" rejected above.

Take *Archaeopteryx* for example (Rieppel, 1984d), a fossil hailed as perfect intermediate between reptiles and birds! From a pattern point of view (Gauthier and Padian, 1985), *Archaeopteryx* constitutes the sister-group (Archaeornithes) of all higher birds (Neornithes): the fossil shares with higher birds some, but not all of their diagnostic features. If *Archaeopteryx* is to be the actual ancestor of higher birds (Ostrom, 1976), it must not have diagnostic characters of its own, or reversal would have to be assumed for these characters if the transition from *Archaeopteryx* to higher birds is to occur. If *Archaeopteryx* has no diagnostic homology of its own (autapomorphy), there is no way to know whether the fossils referred to the genus and its only known species, *Archaeopteryx lithographica*, do indeed form an unequivocal group which could be interpreted as a species or population ancestral to higher birds. It has indeed been argued that there existed two genera and species of Jurassic birds in the Solnhofen lithographic limestones (Howgate, 1985). As Darwin equated evolution with the origin of new species, it can no longer be argued that *Archaeopteryx lithographica* is *the* ancestor of higher birds, since we do not know what *Archaeopteryx lithographica* in fact is. Patterson and Rosen (1977) have formalized the way to classify fossil groups of paraphyletic status with respect to monophyletic extant taxa by the introduction of the plesion concept. The plesion is a fossil taxon "sequenced in a classification according to the convention that each such group is the (plesiomorph) sister-group of all those, living and fossil, that succeed it" (Patterson and Rosen, 1977: 160).

We might find out what *Archaeopteryx lithographica* is, if it could be demonstrated to represent an unequivocal group. Indeed, diagnostic characters of the genus were already recorded by Ostrom (1976), amongst which the shape of the distally bifurcated ischium features prominently. A recent re-investigation of the skull has produced further autapomorphies of *Archaeopteryx*, including the morphology of the quadrate bone and the shape of the squamosal, if this element (retained in higher birds) was at all present (Whetstone, 1983). These characters (where preserved in the fossils) diagnose an unequivocal group, but if that group is made the ancestor of higher birds, reversal must be assumed for the features in question, since they are absent in higher birds. The taxon *Archaeopteryx*, in a terminal position on the cladogram, is forced down on a branching point (node) of the cladogram in order to provide the actual ancestor for its sister-group, but by this standard ancestry becomes a problem of likelihood of character reversal.

The cladogram specifies sister-group relationships on the basis of homologies shared at the correct level of inclusiveness (synapomorphies). These homologies characterize a hierarchy of types within types, the type representing an abstract notion of logical relations of similarity. The type is not an actual organism, let alone an actual ancestor: the type is an inclusive taxon, and a guide to common ancestry if viewed from an evolutionary perspective. Darwinian evolution, on the other hand, requires that taxa stand in a relation of ancestors and desendants to one another. This relation can be satisfied by exclusive and paraphyletic taxa only: a reptile which is ancestral to birds cannot itself be a bird (although reptiles and birds can have a common ancestry if they fall into a common group of greater inclusiveness, the Sauropsida). Exclusive taxa are paraphyletic if they have no unique character of their own, i.e. if they do not form unequivocal groups. In other words, the taxa which are to be ancestral or descendant, cannot be unequivocally objectified! The conclusion must be that logically stringent pattern analysis is, indeed, incompatible with the Darwinian theory of evolution (Nelson and Platnick, 1984).

The hierarchy of homologies, a logical conception of biotic diversity, ahistorical and acausal in nature, specifying inclusive taxa and therewith emphasizing discontinuity between groups, is incompatible with the Darwinian process of evolution. But on the other hand, each subordinated hierarchy of groups within groups is complementary to a serial arrangement (admitted to include branching points or bifurcations) of the exclusive taxa which it comprises. This complementarity opens the path to a causal explanation of the cladogram, which therewith changes from an atemporal and ideal scheme of logical relations to a temporalized expression of contingent historical relations.

There are two possible conventions to transform a static cladogram into the temporalized version of a phylogenetic tree. One possibility is to mentally add the implication of ancestry to branching points: "This makes what has been called an 'X-tree', a tree containing no ancestors (each ancestor is an unknown 'X') but carrying the implication of time-scale and of descent and speciation" (Patterson, 1983: 26). The "X-tree" corresponds to a "minimal Steiner tree" in graph theory (Page, 1987: 5) with no actual data at the nodes. The latter represent nothing but generalizations about some hypothetical common ancestor derived from the synapomorphy scheme determining the relations among the terminal taxa. The notion of ancestry implied in an 'X-tree' is that of common descent from a *hypothetical* ancestor of all those forms belonging to the same type or subtype. With reference to the example discussed above it would be stated that

Archaeopteryx and higher birds share a common ancestry, or are descended from a common hypothetical ancestor. This type of causal explanation of the hierarchy of homologies requires no compatibility with the notion of actual ancestry, but it imposes a time-scale on the cladogram, and postulates that the discontinuities on which the hierarchy is based must be bridged by some continuous process.

The alternative convention is to force one or several terminal taxa down on branching points (nodes) of the cladogram in order to provide an actual ancestor(s) of the group (type) characterized at that node. By this procedure, part or the whole of the cladogram becomes translated into a serial arrangement of forms, one succeeding the other. Again referring to the example above, and assuming that *Archaeopteryx* has been demonstrated to be the sister-group of higher birds, the procedure would be to force *Archaeopteryx* down on the node at which it and the higher birds pair off, in order to provide the actual ancestor for the latter group. The adequacy of this procedure depends on the degree of resolution which has been obtained in cladogram reconstruction.

Complete resolution is achieved in pattern analysis when all the groups recognized are unequivocal, i.e. when all sister-group relations are defined on an unequivocal basis. This not only implies the diagnosis of types by homologies at their correct level of inclusiveness, but it also requires the diagnosis of terminal taxa as unequivocal, i.e. individualized groups, characterized by autapomorphies. Consider a simple three-taxon statement, involving amphibians, reptiles and birds. Assume these three taxa to belong to a group of higher rank, the Tetrapoda, characterized by homology $\underline{A}$ in Fig. 17; assume further that reptiles and birds are more closely related to each other than either is to amphibians on the basis of character $\underline{B}$ in Fig. 17, so that reptiles and birds are included within the Sauropsida, the latter subordinated to the Tetrapoda. The following classification obtains: (amphibians (reptiles, birds)). In that classification, two types of taxa must be recognized, viz. exclusive versus inclusive ones. The terminal taxon "amphibians" is exclusive: it comprises all tetrapods which are not sauropsids. The "reptiles" include all sauropsids which are not birds. In contrast, the "Tetrapoda" and "Sauropsida" are inclusive taxa: tetrapods include amphibians, reptiles and birds. Nelson and Platnick (1984) note the complementarity of exclusive and inclusive taxa, reflecting the complementarity of the serial versus the subordinative arrangement of groups: amphibians (exclusive taxon) comprise all tetrapods which are not sauropsids; the Sauropsida (inclusive taxon) include all tetrapods which are not amphibians. There is an important difference between these taxa, however. The amphibians *as characterized above* are a privative group and hence paraphyletic. Amphibians have so far only been characterized by tetrapod features which do not segregate them from reptiles and birds, or conversely: they have so far been characterized by the absence of sauropsid characters only. Similarly, all characters used to diagnose reptiles (sauropsids which are not birds) are plesiomorphic, i.e. diagnostic at a more inclusive level, characterizing the Sauropsida. The inclusive taxa Tetrapoda and Sauropsida, on the other hand, are unequivocal groups, diagnosed by the characters $\underline{A}$ and $\underline{B}$, respectively (in Fig. 17). What is the implication of paraphyly and monophyly respectively if a terminal taxon is forced down on a node in the cladogram to provide an actual ancestor?

The classification as discussed above (and implied in Fig. 17) suggests that the amphibians are the sister-group of the Sauropsida. The cladogram furthermore restricts the ancestry of the Sauropsida to amphibians (Nelson and Platnick, 1984). Does this mean that amphibians can be forced down on the node $\underline{A}$ to provide the ancestor for the Sauropsida? As a matter of fact, this conclusion does not necessarily follow at this stage of analysis. All that can be stated is that reptiles and birds are more closely related to each other than either is to amphibians, due to the character ($\underline{B}$ in Fig. 17) shared by reptiles and birds but not by amphibians, characters which therefore diagnose the Sauropsida. Since all three terminal taxa belong to the Tetrapoda (by virtue of character $\underline{A}$ in Fig. 17), it can furthermore be stated that the sister-group of the Sauropsida must be sought within those tetrapods which are not sauropsidans, i.e. within amphibians. Since amphibians have not yet been diagnosed as an unequivocal group by a uniquely shared homology (autapomorphy), it is not known what amphibians really are. Since the amphibians form no unequivocal group, they cannot constitute the sister-group of Sauropsida - they can only comprise this sister-group, which indeed must be the case if the Tetrapoda is a valid taxon. This is in accordance with the requirement that higher taxa cannot represent ancestors: if anything, only species, or rather populations, can be considered as actual ancestors. The "amphibians" therefore cannot be ancestral to sauropsidans, but they must contain some species, or some subgroup of a species, which would be closest to the Sauropsida, which therefore would be their sister-group, and would thus provide the actual ancestor.

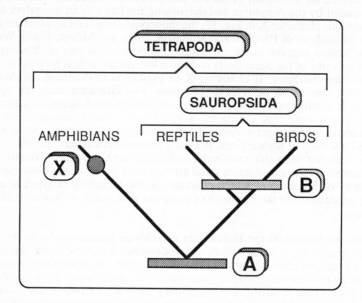

Fig. 17: A cladogram of tetrapods and its relation to conventions of ancestry. For further details see text.

According to Lovtrup (1979: 387), "...a taxon becomes defined by the possession of the characters x, y, z, etc., if, and only if, it has a twin or sister taxon the members of which share with it all characters except these". This convention specifies the definition of inclusive taxa as one proceeds along the lower edge (axis) of the cladogram. Resolution of pattern analysis remains incomplete, because all terminal taxa (the "twin" or "sister" taxa of the above quotation) are privative or paraphyletic groups. In this case, nothing precludes forcing down any terminal taxon on a node in the cladogram to provide the actual ancestor. On the definition given above, nothing precludes amphibians to be the ancestors of sauropsidans, but at the same time it must be acknowledged that we do not really know what amphibians are: the ancestral taxon has not unequivocally been diagnosed as a group. It contains the sister-group of the taxon at the next lower level of inclusiveness, but it does not necessarily represent it. The same probem obtains with respect to the ancestry of birds: as long as *Archaeopteryx* shares no unique homology, nothing precludes it from being ancestral to birds, but at the same time we do not know what *Archaeopteryx* really is: is it an ancestral species or a paraphyletic genus? The quintessence is that a cladogram in which the inclusive taxa alone are unambiguously characterized, whereas the terminal taxa remain paraphyletic, is directly convertible into a serial arrangement, i.e. into a phylogenetic tree. It is possible, in the course of the search for the sister-group of the Sauropsida within the paraphyletic amphibians, to continuously subdivide the latter assemblage into paraphyletic groups of lesser and lesser inclusiveness, sharing an ever-increased number of characters with the Sauropsida and thus coming closer and closer to the species supposed to be ancestral to reptiles (Fig. 18). The paraphyletic Amphibia thus dissolve into the continuous stem-lineage (Ax, 1984) of the Sauropsida, along which the characters diagnosing the Sauropsida are gradually evolved (for further discussion of this point see below). Continuity of change has thus been imposed on the cladogram by its conversion into a linear arrangement of forms.

The degree of resolution in pattern analysis is increased as a homology is found uniquely shared by the Amphibia and diagnosing the latter as an unequivocal group (the autapomorphous character $\underline{X}$ in Fig. 17; the monophyly of extant amphibians is supported by the recent analysis of Panchen and Smithson, 1988, and Milner, 1988). We then know what Amphibians are, but at the same time the position of one of their representative species as ancestor of the Sauropsida is thrown into doubt, unless reversal is assumed for that amphibian character. If character $\underline{X}$ in Fig. 17 is to diagnose the Amphibia as a monophyletic, i.e. as an individualized group, this character must not appear in the Sauropsida. If amphibians are to provide the ancestor for the Sauropsida, character $\underline{X}$ must have been lost at some time during this process of descent. This is the same problem as discussed above with respect to the unique features of *Archaeopteryx*, ruling the fossil out from direct ancestry of higher birds. If a fully resolved cladogram, in which not only the inclusive taxa but also the terminal (exclusive) taxa are unequivocally diagnosed (individualized), is translated into a serial arrangement of forms, the latter can only mirror the hierarchy of homologies which characterize the hierarchy of types in a downward direction; information on the diagnosis of exclusive taxa is lost, or incompatible with serial arrangement.

Again, we encounter the Haeckelian versus von Baerian conventions of ancestry. Haeckelian groups are actual ancestors and descendants respectively, and thus by definition privative or paraphyletic. As compared to the ancestor, the descendant is derived by the addition of an evolutionary novelty: it shares with its ancestor all the latter's diagnostic features, plus some additional evolutionary novelty absent in the ancestor. As a consequence the ancestor must be a privative or paraphyletic group with respect to its descendant. Von Baerian groups on the other hand are types, logical conceptions of similarity providing a guide to common ancestry. What can be translated into a linear sequence is the subordination of types, i.e. the generality of character distribution as indicated along the lower edge (axis) of the cladogram: this is preserved as trunk in the convention of the phylogenetic tree. The real organisms sit out on the branches, and cannot be ancestral one to another. Tetrapods cannot be the ancestor of mammals, although the ancestor of mammals must have been a tetrapod, and all mammals are tetrapods: the shared presence of tetrapod characteristics is a guide to common ancestry of mammals and some other tetrapod, but it does not specify an actual ancestor. As Patterson (1983: 12) put it, the Haeckelian convention of ancestry is one where "a group gives rise to another one of equal or higher rank, and the pattern of character distribution is that the ancestral group is always characterized only by the lack of the features which distinguish the descendant group, i.e. by absent or privative characters". The von Baerian convention of ancestry, on the other hand, "includes groups whose rank decreases as one ascends the tree and unknown or hypothetical members of such groups", and hence is one of "decreasing generality" (Patterson, 1983: 12).

It might be argued, of course, that all of these considerations apply only in cases of supraspecific groups unduly made ancestors. Problems dissolve at the species level, because species *qua* ancestors need not qualify as monophyletic, i.e. need not be unequivocally diagnosed by a uniquely shared character. The same point is entailed in the asertion that species *qua* individuals need not be defined, but can only be recognized

(Ghiselin, 1974). This point has been dealt with before, under the heading of the paradox of the evolving species (chapter 2), and it may be further elaborated with reference to Fig. 18. Species have been called the "smallest detected sample of self-perpetuating organisms" (Nelson and Platnick, 1981: 12). In order to detect a smaller sample within a larger sample, no uniquely shared character of the smaller sample is necessary. Assume a larger sample of organisms, diagnosed as an unequivocal group on the basis of some homology (character x), and including the smaller samples A and B. Assume that within the larger sample, the smaller sample B shares a homology (y) at a less inclusive level, and is thus recognized as an unequivocal subgroup. The initial sample has thus been subdivided into two smaller samples, one containing taxon A, the other containing the taxon B; no further subdivision is possible. By these standards, taxon A is not unequivocally diagnosed, but it can be recognized: it comprises all organisms sharing character x but lacking character y (see Lovtrup's convention above). If such is deemed acceptable, a standard has been created to recognize what Ax (1984) has called a stem-species. The implication is that if an evolutionary lineage undergoes dichotomous speciation, one of the resulting sister-species must carry an evolutionary novelty (to objectify the splitting event), while the other species may remain unmodified and in that case can be thought of as representing the actual ancestor, i.e. the stem-species of the dichotomy in question. After all, some species must have been ancestral if evolution has occurred!

Ax (1984) has argued that any higher taxon, for instance the Mammalia, is likely to be diagnosed not only by one, but rather by several homologies. He further argues that it is unlikely that all these diagnostic features should have evolved at once, but rather have developed in succession, as is documented by the lineage of mammal-like reptiles. To account for these considerations he introduced the concept of the stem-lineage, leading from the reptilian ancestry up to mammals through a series of stem-species, each characterizing a node at which dichotomous speciation created a new mammalian homology. It is obvious that the stem-lineage concept of Ax (1984) represents a phylogenetic interpretation of the plesion concept proposed by Patterson and Rosen (1977). The sequence of fossils as plesiomorph sistergroups of all those which are to follow along the axis (lower edge) of the cladogram corresponds to the sequence of stem-species along the stem-lineage.

The argument saves the notion of actual ancestry by denying the necessity to characterize species as unequivocal (monophyletic) groups. The consequence, however, is that the stem-species, or an ancestor at species level, must necessarily be a privative or paraphyletic group. The notion of ancestry is saved, but it cannot be objectified! If we objectify the ancestral or stem-species as a privative group, we cannot know whether it really represents the ancestor: we can only assume that the group supposed to represent the stem-species comprises the ancestor. With reference to Fig. 18 it may be assumed that taxon B is characterized by a suite of characters all absent in taxon A. Assume these taxa to represent species of an unequivocal (monophyletic) group of higher rank. In this case, A could be interpreted as stem-species of the lineage leading up to B. Is A therewith objectified as ancestor of B? All that can be stated is that A contains the ancestor of B. Since A is privative, a search for the actual ancestor of B may result in a successive subdivision of A, the subgroups approaching B evermore closely, sharing an ever-greater number of characters typical of B. As long as the subgroups of $A(A', A''. A''')$ remain privative in their turn, subdivision may eventually continue down to individual organisms.

The supposedly ancestral species has gone lost; it has become subdivided into subgroups of progressively lesser inclusiveness which progressively approach type B. In essence, the question has become one of cladistic relations of populations or demes (Brady, 1983; Schoch, 1986), or, in the extreme, of individual organisms. The search for the actual ancestor within species A has led to the concept that some demes (or organisms) of A are closer to B than others. We simply slide down the genealogical hierarchy in the search for an actual ancestor within a privative group. The analysis of ancestry comes to an end with the elucidation of cladistic relations between species and/or populations, each characterized as an unequivocal group by unique characters. In this case it becomes possible to unequivocally objectify sister-group relationships, but the notion of actual ancestry is bound to character reversal.

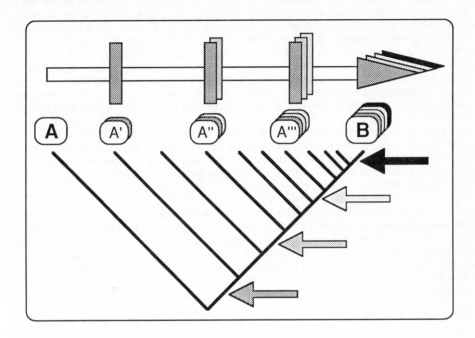

Fig. 18. The relation of the stem-linage concept to a subordinated hierarchy of groups within groups.

Comparing the genealogical with the ecological hierarchy, we see "demes" and "populations" as virtually interchangeable notions: what else is a deme but a localized population? That we should slide down the genealogical hierarchy in the search for actual ancestors may simply be another expression of the fact that species do not represent the basic unit of evolution. Instead, evolutionary change takes place within populations, embedded in their local ecosystem. One might argue that this only shifts the problem of the definition or diagnosis of an unequivocal group down the hierarchy. Not necessarily so, if demes and populations are accepted as corresponding concepts. Demes, or localized populations, are the hierarchical level at which *actual* interbreeding can be ascertained. Reproduction is the *sine qua non* for evolutionary change (at least in most metazoans). The actual ancestor therefore has to be an interbreeding group of organisms. If we take this to be the species, the problem of *actual* versus *potential* interbreeding remains as the species extends through time and space. This is the reason why species as entities of the genealogical hierarchy, extending through time and space, can and must be diagnosed as unequivocal groups, whereas the notion of actual ancestry must be restricted to localized, i.e. actually interbreeding populations or demes. However, the study of such populations will only provide us with the clues to the mechanisms of evolutionary change, but not with the clues to the pattern of the ancestral and descendant condition, i.e. with the clue to the direction of change. This problem originated from the incompatibility but complementarity of process versus pattern analysis: process analysis provides the mechanisms of change, but only pattern analysis can provide a direction for change. Only process analysis in localized populations can accommodate the notion of actual ancestry, but only pattern analysis can polarize the hierarchy of types.

In summary it can be stated that the analysis of ancestry does not differ, in essence, if carried out on the species level or on supraspecific levels of the genealogical hierarchy. In either case the genealogical hierarchy must first be objectified by the generality of homologies: a hierarchy of groups within groups obtains, whereby some level of generality may or may not be interpreted as the species level with reference to some biological species definition (for instance the level at which interbreeding still takes place). The notion of ancestry presupposes the reconstruction of the hierarchy of groups within groups to provide the direction of evolutionary change, but the study of mechanisms of evolutionary change relates to the level of populations as part of the ecological hierarchy. The notion of actual ancestry, and with it the continuity of the process of descent with modification, can only be imposed on a hierarchy of groups within groups if the terminal taxa remain privative. As the terminal taxa become unequivocally objectified as groups on the basis of some uniquely shared homology (autapomorphy), the notion of actual ancestry must imply character reversal if any of these terminal taxa is forced down on its node in the cladogram in order to serve as actual ancestor.

If we admit privative species, we objectify the species taxon as an open system and we obtain evolutionary lineages. If we close the species taxon by its objectification (i.e. individualization) on the basis of some uniquely shared character(s), species can no longer serve as actual ancestors unless reversal for those diagnostic characters is assumed (Rieppel, 1986b).

The Notion of Convergence

If the hierarchy of groups within groups is explained on the hypothesis of descent with modification, the characters diagnosing groups at different levels of inclusiveness are interpreted as evolutionary novelties, "invented" by the immediate ancestor of that group (Rieppel, 1983a): The jaws are an evolutionary novelty having arisen within the Vertebrata; the tetrapod limb is an evolutionary novelty characterizing a subgroup of the Gnathostomata; the amnion is an evolutionary novelty having arisen within the Tetrapoda.

On the theory of evolution, the relation of homology becomes a relation of descent: the relation of coexistence of structures becomes interpreted as a relation of succession of structures, similarity becomes a guide to common ancestry. The Amniota are more closely related to each other than any amniote is to any anamnian tetrapod, because the Amniota share a common ancestor (the 'creator' of the amnion) which is not also the ancestor of anamnian tetrapods. By this interpretation of homologies, no actual ancestor is evoked: the level of generality at which a homology forms a group is a guide to common ancestry. None of the difficulties incurred with Haeckelian ancestors obtains. Shared homologies are explained by the inheritance of ancestral structures. Inheritance of these ancestral structures may be explained by their membership in a causal ontogenetic sequence (structuralism) or by the concept of adaptation (functionalism).

As was discussed above (chapter 3), a useful (maximally congruent) hypothesis of grouping has no true predictive power, since it does not address reproducible processes. Nevertheless, the prediction of the criteria of membership in a group (of the homologies characterizing that group) is sometimes highly successful. Should any such prediction go wrong, however, the organism in question is either excluded from the group, or the diagnosis of the group is revised so as to accommodate the organism in question. Predictions of similarity based on classification turn out to represent rationalizations after the fact of pattern reconstruction. It is pattern analysis which indicates a greater or lesser degree of congruence for different characters, and these become only predictable on the basis of a causal theory which explains their congruent distribution, or the lack of congruence.

An interesting observation common in pattern analysis is that incongruence does not affect every hypothesis of grouping and every kind of character to the same extent. Some hypotheses of grouping are very stable: the Metazoa, the Echinodermata, the Vertebrata, the Gnathostomata, the Amniota, the Aves are successful hypotheses of grouping which appear to support the notion of congruent order in nature. Every child readily distinguishes a butterfly from a bird! Everyone would probably be quite confident to be able to recognize a mammal, since all mammals have hair, mammary glands, three ear ossicles etc., and the prediction of these homologies for any mammal is unlikely to be wrong (Riedl, 1975). However, when all the extinct mammal-like reptiles, the Mesozoic mammals, and the monotremes are included in the analysis of a proper delineation of the taxon Mammalia from non-mammalian amniotes, the assessment of the proper level of generality for those features diagnosing the Mammalia becomes problematical, and incongruent distribution of osteological features adds to this difficulty.

Congruence appears to be high for those characters diagnosing more inclusive groups at high levels in the genealogical hierarchy. As one moves down to lower, less inclusive hierarchical levels, the amount of incongruence increases, and the prediction of homologies becomes risky and frequently wrong. The analysis of pattern among chamaeleonid genera using skull characters resulted in such a high degree of incongruence that the choice among alternative hypotheses of grouping might seem to boil down to a matter of taste (Rieppel, 1987c). The head morphology of acontine skinks (Rieppel, 1982) and the classification of primitive snakes (Rieppel, 1979a) are other examples of incongruence at the level of families and genera. The problem of incongruence becomes acute as one proceeds downward to the level of reproducing communities. At the level of species and below, incongruence obtains to such an extent that it has been questioned whether the methods of pattern analysis outlined above can at all work at these low levels (Arnold, 1981). This is not to say that higher groups are never prone to incongruence: the classification of snakes involves incongruent character distribution at the family level and above (Rieppel, 1988a), and the Choanata as proposed by Rosen, Forey, Gardiner and Patterson (1981) is another much debated case in point (Jarvik, 1981; Schultze, 1981; Thomson, 1981; Mee-Mann and Xiaobo, 1984; Holmes 1985)! Multiple and equally parsimonious hypotheses of grouping likewise frequently obtain, even at higher hierarchical levels on the basis of molecular data (Romero-Herrera, Lehmann, Joysey and Friday, 1978).

From the observation of incongruent distribution of topographical correspondences results the problem of how to explain character conflicts, increasing in number with decreasing level of inclusiveness of the groups in question. One possible causal explanation of character congruence versus incongruence was already offered in the preceding discussion of causal ontogenetic sequences (Alberch, 1985). If some character forms part of a causal ontogenetic sequence, its differentiation becomes a *sine qua non* for harmonious development. For developmental reasons the character in question becomes a constituent element of the type characterized by it. The earlier the character in question originates during development, the more profound will its influence be on subsequent stages of the (continuous) process of ontogenetic differentiation. As long as von Baer's first law holds, the character originating early during ontogeny will diagnose a more inclusive group as opposed to features which make a later appearance. Riedl (1975) developed the concept of *burden* to characterize the relation of causal ontogenetic sequences to classification. Features originating early during ontogeny and characterizing more inclusive groups will carry a heavier burden by virtue of their greater significance in a causal ontogenetic sequence. The heavy burden explains the highly congruent behavior of these features, and renders their prediction successful within the limits of von Baer's law. Features originating later during ontogeny and carrying a lesser burden may behave less congruently and their prediction is more likely to be wrong. For characters which do not form part of a causal ontogenetic sequence there is no reason to expect congruent distribution: they are useless for systematic purposes (Alberch, 1985), as they do not carry any burden.

Burden of a character, imposed on it by its relation within a causal ontogenetic sequence, thus becomes a measure of its phylogenetic information content. The phylogenetic information is expressed by the congruent distribution of the characters in question. This is just another way to say that characters weight themselves - by congruence (Patterson, 1982).

However, the first law of von Baer may not hold (as in the case of the Amniota), or a causal ontogenetic sequence may not be conservative: it may change in the course of evolution. Von Baerian recapitulation and the related concept of burden have a limited predictive power (Alberch, 1985; Shubin and Alberch, 1986). Kluge and Strauss (1985:164) note that there is no reason to expect greater conservatism in developmental pathways than in conventional taxonomic characters. The first step, as always, has to be the reconstruction of taxic relations. On this basis it would seem possible to explain a high degree of character congruence on the hypothesis of a heavy burden, a hypothesis which might even become subject to some experimental test of the developmental and/or functional integration and correlation of the characters in question.

Arthur (1984) developed a largely hypothetical (Forey, 1985; Hanken, 1986) model of genetic control over the epigenetic process of ontogenetic development. The von Baerian hierarchy of character differentiation results in a subordinated hierarchy of progressively less inclusive levels of generality. This hierarchy of groups within groups is mirrored, according to Arthur (1984), in a hierarchy of "developmental genes" (abbreviated as "D-genes") which control development. The "D-gene" is a complex concept involving a structural gene controlled by a regulatory gene and coding for an enzyme whose product is a morphogene, a substance which determines processes of morphogenesis (Arthur, 1984: 149). The hierarchy of "D-genes" would be the mirror image of the genealogical hierarchy. Characters of a high degree of generality, diagnosing groups of great inclusiveness and ranked towards the base of the genealogical hierarchy would be determined by one or few "D-genes" active during early phases of morphogenesis. Characters of lesser generality, diagnosing groups at a less inclusive level of the genealogical hierarchy, would be determined by the interacting effects of a large number of "D-genes" switched on during later stages of ontogenetic differentiation. The corollary is that those few "D-genes" active during early phases of morphogenesis determine the basic features and their mutual relationships which characterize a more inclusive group. Any mutation of such a gene would consequently be of far reaching effect and very likely to "disrupt a delicately balanced and highly integrated system" (Gould, 1984: 237) rather than to produce a viable evolutionary novelty. This would explain the conservatism of early ontogenetic stages and therewith the high degree of congruence among the characters involved.

"Adaptive modifications are far more likely to affect the later, specific stages of ontogeny" (Gould, 1984: 237), i.e. characters which, according to Arthur (1984) would be programmed by a large number of "D-genes" switched on during later stages of development and interacting in the sense of polygeny and pleiotropy. This allows not only for mosaic evolution, but also for mutations which do not threaten to disrupt the basic organization of the whole organism but rather may serve as raw material for adaptation by natural selection. Mosaic evolution and convergent adaptations are thus invoked to explain the high degree of incongruence often observed in characters of lesser generality.

The Complementarity of Pattern and Process

Pattern analysis is based on the relation of homology, i.e. on logical and hence static relations of similarity which abstract from specific form and function. The method of pattern reconstruction is downward classification by logical subdivision. Based on analytical thinking, the method seeks to oppose one group to the other by the determination of alternative characters or character-states. The "pattern way of seeing" is rooted in an essentialist tradition. For Aristotle, the species was characterized by its *differentia* from other species within the *genus proximum*. The *genus proximum* provides the higher level hypothesis and therewith the frame for the diagnosis of one subordinated group as opposed to the other on the basis of essential *differentiae*, of diagnostic features. Diagnostic features provide the criteria of membership in a group which by these characters is clearly demarcated from other such groups. There is no transition from one group to the other, no transformation of one type into another.

Pattern reconstruction is in the first place a procedure of information storage, of classification of groups within groups in the most parsimonious manner so as to maximize congruence. Classification emphasizes discontinuity and the subordinated hierarchy of types and subtypes. The type is a logical construct, a group diagnosed by homologies at their correct level of inclusiveness. The name designating the type carries maximal information on that group and renders communication about entities in nature optimal. By its logical construction, the hierarchy of types is static, i.e. ahistorical. Homolgies denote *relations of coexistence* of constituent elements in organic structures (Rieppel, 1985b). By its abstraction from specific form and function, the hierarchy of types is acausal: it abstracts from the causes (structural or functional) of similarity versus dissimilarity and change, but remains restricted to the representation of formal, i.e. topological relations of similarity. The type represents an element of 'being' in a process of continuous actualization of form. As such, the type is a problem not of causality, but of *Anschauung* (Rieppel, 1985c). Focusing on the invariance of topological relations "independent of their historical appearance" (Goodwin, 1982: 51), it assumes a regulative function, guiding observation in the attempt to conceptualize a unity of organic appearances (Lenoirt, 1978). As Goethe emphasized: the *idea* of the unity of type must rule observation in order to transcend mere multiplicity, reducing it to hierarchical order. The call for causes which would explain this hierarchy of type is a different problem (Brady, 1983, 1985), and it must follow the reconstruction of the hierarchy of homologies.

By its causal explanation, the hierarchy of types becomes submitted to the constraints of time and space. Structuralism addresses the causes of invariance of structural relations in the process of actualization of form within a field of developmental potentialities. If freed from evolutionary conceptions, it preserves the original meaning of homology and therewith the relations of structural coexistence specified by the hierarchy of types (Goodwin, 1982; see also Russell, 1916). Functionalism on the other hand focuses on the determination of form and function by external conditions of life, and thus addresses the contingencies of the evolutionary process.

The hierarchy of types is incompatible with but complementary to the linear sequence of exclusive taxa which it comprises (Nelson and Platnick, 1984). Forcing terminal taxa down on their nodes in the cladogram transforms segments of the latter into a linear sequence of forms amenable to evolutionary explanation, specifying ancestral and descendant conditions. The static hierarchy of types is thus transformed into a temporal sequence of form: homology thereby becomes a *relation* of *succession*, a relation which is explained by the inheritance of adaptations (functionalism) or of developmental constraints (structuralism) from a common ancestor.

The causal explanation of the cladogram by the hypothesis of descent with modification transforms the latter into a phylogenetic tree. Whereas the cladogram is in itself a logical conception and therefore acausal and atemporal, the phylogenetic tree is built on causality and historical contingency. On the one hand, the cladogram is incompatible with the phylogenetic tree, on the other hand the two conceptions are complementary to each other. The method of pattern analysis is incompatible with the evolutionary research program, because pattern analysis strives for the diagnosis of unequivocal groups (exclusive taxa being individualized by diagnostic features of their own, i.e. by autapomorphies), whereas only privative (paraphyletic) groups can be forced down from a terminal position on a node in the cladogram to provide an actual ancestor (Nelson and Platnick, 1984). By this procedure, however, the actual ancestor dissolves into a string of evolving populations.

The "process way of seeing" may thus lead to a nominalistic perspective. Related to an atomistic conception of ontogeny and organization, it offers a theory which renders phenotypic variation fundamental and therewith all possible intermediate morphologies at least potentially possible. The gaps in a discontinuous series of forms are bridged on the hypothesis of a continuous series of intermediates, what renders all demarcation of groups subjective or arbitrary.

In contrast, structuralism is related to an epigenetic conception of ontogeny, offering a theory of continuous causes to explain discontinuous adult morphologies: transformation at the phenotypic level appears saltatory, although it is the effect of continuous genetic and epigenetic causes.

The notion of actual ancestry is incompatible with a logically stringent pattern analysis. The way out of this dilemma is the 'X'-tree, based on the addition of the time-axis and of the notion of the "hypothetical common ancestry" to the cladogram. This solution is based on the complementarity of pattern versus process: as the axis (lower edge) of the cladogram corresponds to the serial arrangement of the exclusive taxa which it comprises, it indicates the direction of evolutionary change. The axis (lower edge) of the cladogram which defines the subordination of types indicates the direction of palingenesis, while the individuation of terminal taxa indicates the direction of divergent evolution. Only the hypothesis of a phylogenetic process can provide a natural causal explanation for order in nature, yet the direction of this phylogenetic process must be indicated by independent pattern analysis. Every hypothesis of transformation presupposes a hypothesis of relationship.

CHAPTER 7

A MATTER OF LAW

The logical way to reconstruct the hierarchy of homologies is by the method of outgroup comparison or by the comparison of the ontogenetic process of differentiation. Both methods serve to assess the level of generality at which a character diagnoses an unequivocal group, and both are based on the expectation, i.e. on the initial hypothesis, that homologies can be ordered hierarchically (Brady, 1983: 51). In other words: the attempt to reconstruct the hierarchy of homologies is a matter of "order versus disorder " (Patterson, 1982: 42). The real problem, therefore, is the incongruent distribution of characters which have passed the test of similarity. If incongruence obtains, the principle of parsimony must guide the selection of the maximally congruent hierarchy within the given data set (Brady, 1983): among the several possible hypotheses of grouping, the one is selected which shows the greatest degree of regularity in character distribution. "The purpose of science...is to discover and explain regularities in nature, and if we give up the search for regularities, we also give up the game of science. So we may assume that order really does exist and that we, and not nature, are responsible for the apparent disorder", write Nelson and Platnick (1981: 23). If the hypothesis of order in nature is accepted in the context of empiricism, it generates the prediction "that there is only one real or natural hierarchy, one correct or true classification" (Patterson, 1983: 17). To be empirical, such a prediction of order in nature (groups fall within groups in a regular pattern) would have to be tested, and the test would be provided by the reconstruction of the hierarchy of homologies. If all the characters analyzed fall congruently into a regular pattern, "we have overwhelming evidence of order" (Patterson, 1982: 42). If, however, incongruent character distributions regularly continue to creep up, the hypothesis of order in nature would seem to fail the test. Really?

> "To the extent that there is order in nature, to the extent that existing classification is an accurate hypothesis about that order, and to the extent that our hypotheses about properties and their distributions among organisms are correct, such predictions [of homologies] will be successful. If they are not successful, we have discovered an interesting problem, either in our perceptions of properties and their distribution, in our classification, or in nature" (Nelson and Platnick, 1981: 9).

What is a Natural Law?

Natural science may be characterized as the search for regularities in the appearance of the material world: "If I were to deny the existence of regularities in nature, I would not be a scientist", writes Mayr (1987b: 215). If these regularities are regular enough, they become the object of natural laws. If the stone which I picked up regularly falls to the earth, no matter how many times the experiment is performed, a lawful regularity is observed calling for a causal explanation which is provided by the law of gravity (see Chapter 1). In proposing this law, Newton made the interesting distinction of an *essential* property of matter versus a hypothetical explanation of observed phenomena. Newton was not striving for absolute truth, nor was he calling for absolute reality. For him, the truth, or correctness, of natural laws was not based on intuitive obviousness or on deduction from self-evident principles, as it was for rationalists; instead, a law was, and still is, considered to be true if it satisfactorily explains and successfully predicts observed phenomena, i.e if it addresses causal relations. The concept of natural laws was thus primarily related to repeatable and hence testable phenomena of the physical world, for instance mechanical relations between physical objects or celestial bodies. The most significant aspect of the natural law is its universal range of validity within identical frame conditions, establishing the lasting validity of the findings of natural sciences.

> "If a natural law is true, it is true anywhere and at any time in the universe just as long as the appropriate conditions are met. It may be that all the regularities which scientists currently think are lawful are merely due to contingencies of the formation of the universe. If so, there are no scientific laws, but *to give up the search for lawful regularities in nature is to alter the goals of science radically*" (Hull, 1987: 173; emphasis added).

The above quotation gives a definition of natural science which entails an imperative: the job of the "right thinking" (Rosenberg, 1987: 197) scientist is to search for lawful regularities; his *intention* must be to discover phenomena which behave in a lawful manner!

Observation, however, and with it the search for lawful regularities, cannot be and never is unbiased. In the most general sense, it must be based on some expectation: research must be expected to reveal something of significance, something worth recording, but the significance can only derive from some theoretical context: that is, observation is always theory-laden. Furthermore, as in the case of comparative biology, observation consists in the establishment of relations between appearances (homology is the relation of similarity: see chapter 3), and this must be based on some methodology, i.e. on operational rules. The search for lawful regularities is thus a complex endeavor: it presupposes a theory, a method, and it must be intended. Appearances, however, may differ with different methods and different intentions (a point to be discussed in more detail below), which is why a lawful regularity of phenomena may escape discovery if theory, method or intention are wrong. Writes Hull (1987: 173): "If we insist on dividing up the living world inappropriately, we will not discover the operative causal regularities".

The recent debate on the ontology of the species, from which the above quotations were taken, has extended the search for universal laws into biology, an intention not accepted by all authors (e.g. Mayr, 1987a,b). The argument is that, whereas species may not appear as individuals, this is nonetheless precisely the way they must be construed if natural laws (such as natural selection) are to range over them (Hull, 1987: 171, 176). Or, to recall Eldredge's (1985) verbiage, the "new ontology" *forces* a view on us which corresponds to the intention of the biologist. The intention may change, however, and with it changes the conception, and eventually even the appearance, of species such as in folk taxonomies (Berlin, Breedlove and Raven, 1966; Brown, 1985).

This is not the point at which to return to the problem of species individuality (see chapter 5). Instead, the arguments presented above will be extended to the problem of order in nature, i.e. to the relation of homology which was characterized as "the only law-like hierarchic relationship in biology above the level of the individual organism" (Patterson, 1982: 22).

The Law-like Relation of Homology

The object of comparative biology, the science of systematics in its broadest sense, is the discovery of order in nature, the hierarchy of groups within groups as revealed by the relation of homology. The approach is based on the intention of representing the observed similarities between organisms in a hierarchical structure (Brady, 1983), and on the *a priori* acceptance of methodological rules guiding the conceptualization of homologies. If the relation of homology, determining the hierarchy of groups within groups takes on a law-like regularity, it not only corroborates the initial expectation - that nature is ordered hierarchically - but also does it call for a causal explanation (Brady, 1983). We do not need to know whether groups have a "real" existence "outside the human mind" (Patterson, 1982: 59); the groups exist "whether or not there are any systematists around to perceive and name them" (Wiley, 1981: 72); all we want is a hypothetical causal explanation of observed phenomena: groups within groups - provided these phenomena are regular enough to call for such an explanation (Brady, 1983).

It will be remembered that the establishment of relations of homology presupposes some intention: the intention to relate organisms with each other on the basis of some operational criteria and in a hierarchical manner. The approach remains unquestioned as long as homologies behave in a congruent manner, i.e. as long as the distribution of characters exhibits law-like regularity. What, however, can lawful regularity of order in nature mean if conflicting character distribution, i.e. incongruence of topological correspondence, is recorded?

There are several answers to this question, but they all relate to the basic point made in the preceding paragraph, that all observation is theory-laden and must be intended. With respect to character incongruence it would be possible to change either the initial intention, the initial expectation or theory, or the initial way of looking at nature.

It would seem possible to give up the initial intention, i.e. the search for natural (universal) laws in the context of comparative biology. Mayr (1987b: 215) for instance insists that "To me it seems that the statistically calculated regularities found in living nature are not the same thing as the laws of physics..." and continues: "Popper's falsification principle was in large part based on the universality of laws" (Mayr, 1987b: 213), thereby touching upon an important point. In a strict sense, Popper's falsificationism in its early (original) version can only be applied to theories permitting reproducible predictions which can potentially be refuted. Classification does not address reproducible processes, however, and thus cannot be claimed to have true predictive powers: that birds are diagnosed by feathers is not a prediction, but a statement of observed similarity, in fact a convention based on the concept of homology. The predictive powers of ontogenetic mechanisms generating form were discussed above and found to be low (Alberch, 1985). What is usually found in taxonomic research, particularly on relatively low hierarchical levels and using molecular data, is extensive character incongruence. From the many possible hierarchies, the most parsimonious one is selected, but parsimony is not a test. Popper (1976) explicitly replaced the parsimony principle by his falsification principle: the best hypothesis, according to his standards, is not the most parsimonious one, but the most stringently testable one. (The problem of statistical generalizations and their testability was discussed in chapter 1.) The revision of the initial intention permits the admission of incongruence and hence the acceptance of it as part of the observed phenomena which need to be explained.

Another consequence of observed character incongruence might be to consider the initial expectation, i.e. the hypothesis of a strictly dichotomous hierarchy, falsified. The distribution of characters does not consistently corroborate a unique hierarchy of order in nature. It would seem possible to ask the question whether order in nature might fit a different pattern, a pattern which in the face of character incongruence would allow for multiple affinities (reticulation). Such patterns were indeed accepted by pre-evolutionary biologists, one of them being quinarism, popular in the first decades of the 19th century (see chapter 3). The interesting point about quinarism is that it combined the intention to find law-like regularity with a different expectation than prevails in modern biology. The intention of quinarists was to seek for groupings in circles of five, revealing the harmonious and lawful relation of numbers as evidence of the rational plan of Creation. The cultural context was different, favoring creationism and rejecting transformism; the initial hypothesis of grouping was different; but the intention to seek for lawful regularity was dominant then also. The point of this discussion is not to resurrect quinarism (which in turn is falsified by an even larger amount of character incongruence), but merely to illustrate how different cultural and hence theoretical contexts may result in different appearances of natural phenomena. This means that law-like regularity, revealed by empirical observation, and the natural laws based on them, can only be meaningful within a given context. The point of multiple affinities, specified by reticulated cladograms, remains an issue in contemporary biology: the problem is to distinguish between convergence and similarity due to hybridization (Nelson, 1983; Schoch, 1986: 158).

The third alternative in the face of character incongruence is to question the observation, i.e. the conceptualization of incongruent relations of similarity. The strategy is fairly simple, and essentially consists in disregarding incongruently distributed characters. Remember that the intention is to find lawful regularity in the order of nature, and that the initial hypothesis is that nature is ordered hierarchically. This initial hypothesis is corroborated by congruent character distribution, which is why Brady (1985: 119) calls congruence the "criterion of success", indicating an emergent and lawful regularity in the pattern of distribution of homologies. As incongruence creeps up, matters become more complicated. Groups subject to incongruent character distribution appear poorly resolved, a state of affairs that may in the first place be attributed to ignorance: "Our claims to knowledge must rest upon the well resolved groups..." (Brady, 1985: 125).

If the intention is to find lawful regularity, it might be concluded that some mistake has been made in the observation or conceptualization of incongruent characters. Recall Hull's (1987) statement given above, that if we insist on subdividing the world in an erroneous manner, we will never discover natural laws. And more: to give up the search for law-like regularity means to give up the "game of science" (Nelson and Platnick, 1981: 23). In that sense, incongruent characters subdivide the world in a wrong manner, and therefore are wrong characters, useless for the purpose of systematics, i.e. they are not homologies! The intention to find law-like regularities cannot admit character incongruence, and if such is observed, it must mean that we have not yet learnt to correctly look at characters and their distribution. The "game of science" requires that we learn to look at nature in the correct way in order to discover lawful regularity!

Character incongruence is thus primarily taken to indicate that something is wrong with the hypotheses of homology included in the analysis. A reciprocally illuminative character analysis may disclose mistakes in the assessment of the level of generality at which a character forms a group, or it may generate insight into a mistaken conceptualization of a given character (Hennig, 1967; Wiley, 1981). After all, characters (i.e. homologies) must be abstracted from the developmentally and functionally integrated whole and conceptualized according to the principle of connectivity. With this still further possibilities unfold for revisions of interpretation, as the establishment of topological relations presupposes an arbitrarily chosen frame of reference which may be altered, giving rise to new conceptualizations of characters.

Another way to get rid of incongruent character distribution is to simply ignore the characters in question, and to base classifications on nothing but congruent data, i.e. on uniquely shared characteristics (compatibility analysis is nothing but a formalized way to ignore incongruent characters). A critic of the concept of the Choanata as proposed by Rosen, Forey, Gardiner and Patterson (1981) might point at the labyrinthodont infolding of teeth which supports a sister-group relation of early tetrapods with rhipidistian "crossopterygians" rather than with dipnoans (Schultze, 1981; it should be noted, however, that since the appearance of the original work on which this example is based, a new fish has been described, *Diabolichthys* [Chang and Xiabo, 1984], currently treated as a sister-group of the Dipnoi [Forey, 1987] but sharing plicidentine). This character simply does not appear on the synapomorphy scheme proposed by Rosen et al. (1981) for the Choanata. The reasons for the omission of this character can be manifold. For instance it might be argued that initial classification must be based on extant animals rather than on fossils, a point forcefully argued by Rosen *et al.* (1981). The rationale is that extant animals permit more complete access to investigation, permitting the analysis of soft tissue and molecular data, whereas fossils are usually incompletely preserved and hence must be classified in relation to a cladogram that has been erected on the basis of extant forms. The same point also obtains from the simple fact that fossils have to be interpreted in the light of a modern model, because only in that comparative frame can some sense be made of their structure. A historical anecdote may serve as an example. During the Middle Ages the nature of fossil shark teeth, indeed their organic origin, remained unknown. They were named "glossopetrae", and their formation was usually attributed to some plastic virtue of natural forces or to some other such elusive mechanism. It was not until 1669 that their true nature was recognized by the Danish naturalist Nicolaus Stensen (Steno), working in Florence at that time. Steno was studying the anatomy of the head of an extant shark, when he was shown "glossopetrae" which had been found in Malta. His immediate conclusion was that these must represent petrified shark teeth, an interpretation of these fossils that became possible only in the light of an extant model. The question raised by this interpretation was how the solid object formed by a once living marine organism could become inclosed within other solid objects, i.e.within rocks. Steno answered the question in his book *De Solido Intra Solidum Naturaliter Contento Dissertationis Prodromus* (Poulsen and Snorrason, 1986).

Referring back to the problem of tetrapod relationships, another possible line of reasoning takes the fact into account that plicidentine does not exclusively characterize rhipidistians and early tetrapods, but also occurs in the actinopterygian *Lepisosteus*, in ichthyosaurs and in the lizard genus *Varanus* (as well as in *Diabolichthys*). As a consequence of its wide and incongruent distribution, the character forms no group and hence is useless for systematic purposes. Uselessness for systematic purposes means that the character in question fails the test of congruence. To remove the character from the analysis on this basis alone is problematical, however.

It was argued in the preceding chapter (see also chapter 3) that the test of congruence is subordinated to the test of similarity (Fig. 4). This means that congruence does not test topological relations of similarity, and therefore does not differentiate biologically meaningful information against noise; rather, it tests the phylogenetic interpretation of topographical correspondence! There is no doubt that plicidentine passes the test of similarity in those taxa where it occurs. It always consists of a centripetal infolding of the dentine into the pulp cavity (Schultze, 1969). Conceptualized as such it constitutes a character (topographical correspondence) which at least potentially may characterize a group. As an example, labyrinthodont infolding of the dentine was treated by Riess (1986: 122) as a character potentially grouping ichthyosaurs with early tetrapods. To remove the character from consideration therefore amounts to the suppression of potentially meaningful information, a procedure which must be based on either of two strategies. Homology is established on the basis of the test of congruence and parsimony alone, an argument that carries the threat of circularity since any conjecture of similarity can be used to test any other and vice versa without the evocation of operational criteria which determine potentially meaningful similarity against non-homology. The other strategy is to base the hierarchy of types on uniquely shared characters only, an argument which fails to give operational criteria for the recognition of shared similarity and hence carries the threat of tautology: a homology is a character which fits a unique and congruent hierarchy - only those characters which fit a unique and congruent hierarchy are homologies.

Both of these strategies can be saved on a definitional basis. If topological criteria of similarity are used as a guide to homology, problems persist as outlined above. If, however, the notion of homology is strictly defined on the basis of common descent, its law-like relation is saved. Since only maximally congruent relations of similarity are accepted as a guide to common ancestry, while all other similarity is disqualified as the result of convergent evolution, it follows by definition that homologies are all fully congruent: they show law-like regularity in their distribution! This definition of homology (and the argument that follows) begs the question how homology as a guide to common ancestry is to be recognized or conceptualized. It begs the question of operational criteria of homology.

This is the reason why a different point of view is advocated here, based on a clear distinction between observation and explanation (Brady, 1985). Comparative biology is a two-step procedure. A first step in the comparison of organisms must address the problem as to what kind of similarity potentially carries phylogenetic infomation as opposed to noise. This distinction is made on the test of similarity. It may be argued that the topological criteria serve as a guide in the observation of phylogenetically meaningful similarity, or they may be said to provide the test for conjectured similarity, whereby the material basis of this conjecture remains irrelevant (Patterson, 1982). The second step involves the causal explanation of topographical relations of similarity, either by common ancestry, or as a result of convergent evolution. Common ancestry versus convergence is tested by the congruence of topographical correspondence. The resulting explanation is a statement of maximal likelihood rather than a denotion of lawful relations.

Admitting that all observation is theory-laden, we may also concede that it is necessary to look at nature the correct way in order to observe meaningful phenomena. One such guide to observation in comparative biology, revealing meaningful similarity, is the *principe des connexions* of E. Geoffroy Saint-Hilaire. It reveals topographical correspondence as opposed to non-homology (see chapter 3). Only those characters which have passed the test of similarity, i.e. which satisfy the operational criteria of homology, are candidates for use in phylogenetic systematics. Topological relations of similarity are all potential guides to common ancestry. Which characters have to be explained by inheritance from a common ancestor is evaluated on the basis of congruence and parsimony. If topographical correspondences show an incongruent distribution, this is not dismissed as due to ignorance or faulty observation, but rather is considered as a biologically interesting phenomenon calling for a causal explanation.

The initial hypothesis of a hierarchical order of nature is not considered falsified in the true sense of the word by incongruent character distribution. That birds are diagnosed by feathers is an observational statement based on the concept of homology and in that sense a "fact" in the wording of Mayr (1987b: 213; it rather represents a convention), not a prediction. That all birds should share feathers is a prediction which cannot be falsified. If the presence of feathers is a condition for membership in the group called 'birds', and an organism is discovered which resembles birds in other characters but which bears no feathers, it is either not a bird, or it is included within the bird group following a revision of the conditions for membership in that group (Rieppel, 1983a). All one can do is search for the most complete hierarchy within the recorded characters on the basis of parsimony, and diagnose the groups accordingly. The addition of further characters may change either earlier hypotheses of grouping, or earlier diagnoses of groups.

Admitting character incongruence as a problem in pattern reconstruction means to address the question of its cause. If the most parsimonious hierarchy of homologies is explained on the hypothesis of descent with modification, the question must also be raised why incongruent distribution of characters can and does occur. Rather than suppressing data, dismissing incongruence as a problem of ignorance, to be solved by future investigation, or as a result of mistaken conceptualization of characters, it might be useful to ask the question why it turns up so consistently, resisting all conceptual attempts to make it disappear. This question links systematics with the study of developmental and/or functional correlation of characters as outlined above (chaper 6). Congruent characters may be shown to constitute links in a causal ontogenetic sequence, or to form necessary parts in a functionally correlated character complex. Incongruent characters on the other hand may be developmentally and/or functionally independent and coded by a variable genetic basis. Their irregular occurrence will thus not disrupt basic patterns of organization. With this knowledge at hand, we obtain a tool for the true prediction of homologies, albeit of limited power (Alberch, 1985; Kluge and Strauss, 1985; Shubin and Alberch, 1986). It is at this stage of the analysis that classification, reflecting the hierarchical order in nature, is no longer an information storage system only, but assumes predictive powers for those characters which have been observed to behave congruently to a high degree. It is now possible to tell *why* all vertebrate embryos will develop segmentation in the paraxial mesoderm, and if the prediction should fail, it is not the hypothesis of hierarchical order which is falsified, but a causal mechanism of ontogeny.

To be sure, pattern analysis has to precede the investigation of character correlation in developmental and functional terms. Experimental embryology or functional anatomy does not result in hypotheses of hierarchical grouping; these must be based on the analysis of character distribution. Only after pattern reconstruction does it become evident which characters show congruent distribution and which ones do not. Only then is it possible to address relevant characters in developmental and/or functional studies. If it is true that "patterns of ontogenetic development need not be any more 'conservative'…than any other conventional taxonomic character" (Kluge and Strauss, 1985: 264), there is no basis on which to predict the congruence of certain characters on *a priori* grounds, i.e. without the implication of some prior hypothesis of grouping. On the other hand, experimental embryology or functional anatomy may *explain* why some hypotheses of grouping are so stable, i.e. provide an *explanation* for a high degree of *observed* character congruence.

The Failure of Evolutionary Epistemology

If character incongruence is admitted as an observational phenomenon in the process of pattern reconstruction, the task of the systematist is to find out which characters form groups and which ones are due to convergence. In other words: several possible hypotheses of grouping obtain, from which the most likely one has to be chosen, a task which is accomplished by the application of the principle of parsimony, maximizing the degree of congruence. The relation of homology cannot be considered law-like in the sense of consistently corroborating a unique hierarchical pattern of order in nature. Instead it boils down to a most parsimonious explanation of topological relations of similarity in terms of descent with modification. This, however, contradicts the basic tenet or departing point of evolutionary epistemology as developed by Riedl and Kaspar (1980) and Riedl (1985, 1986).

The central declaration of evolutionary epistemology is its claim to have solved philosophical problems which have remained unanswered during the last 25 centuries (Riedl, 1986: 15). In fact, arguments of traditional philosophy continued to revolve around realism versus idealism, induction versus deduction, etc., without coming to any conclusion as to which one of the various schools of thought deserves prevalence over the other. Riedl attempted to solve the arguments by cutting through the Gordian knot, declaring them as futile if viewed from the perspective of evolutionary theory. We do not know and need not know whether the world as we observe it is real or not; what we do know, however, is that we are able to act in this world in a manner which secures survival.

And this, Riedl concludes, must be due to the action of natural selection during the phylogenetic past. On the basis of a hypothetical reality, of hypothetical similarity and of hypothetical causes (Riedl and Kaspar, 1980) human thinking assumes similar causes to bring about comparable effects. There is no need to know whether observed effects are "really" similar, whether the inference of particular causes is correct, i.e. "true", or whether reality is "objective". All that is required is that inferred causes are correct to the extent that they permit successful predictions. These predictions guide everyday life, permitting the human being to interact successfully with some hypothetical reality. Were the inferred causes consistently wrong, they would consistently be falsified by everyday experience, and the evolutionary development of the human lineage would long have come to an end. On the hypothesis of evolution, Riedl and Kaspar (1980) and Riedl (1985, 1986) conclude that natural selection has adapted cognitive powers to the environment to an extent which secures survival. This, however, implies that the human powers of cognition are innate, and that the "subjective" structures of cognition of the human mind are isomorphic with "objective" (or "extra-subjective": Riedl, 1986: 43, 105) reality. Or, to put it in other words: the innate structures of cognition are an *a priori* of human perception of some hypothesized "reality-in-itself".

Evolutionary epistemology has been so extensively criticized that it is impossible to repeat all the arguments which have been raised against Riedl (see Löw, 1983; Spaemann, Koslowski and Löw, 184). Only a few points will be mentioned before concentrating on those aspects of evolutionary epistemology which are of particular interest in the present context.

On the basis of his "hypothetical realism" (Riedl and Kaspar, 1980: 31), Riedl has to deny first causes or last ends as much as "absolute facts" (Riedl, 1985: 109, 132; 1986: 141), but this does not prevent him from treating the theory of evolution by natural selection as precisely such a first cause and absolute fact which can no further be questioned. His defense of natural selection is furthermore empirically empty (Brady, 1982) as it falls short of testability and therefore succumbs to the impossibility of logical justification of inductive generalizations (Lewontin, 1972): the action of natural selection is asserted after the fact of survival (e.g. Riedl, 1986: 102) with no prediction being made which would be testable and potentially falsifiable (Rieppel, 1987d). His epistemology finally succumbs to the problem of induction (see chapter 1, and Rieppel, 1986d). According to his argument, causes or explanatory theories are inductively inferred from the observed uniformity of effects; the explanatory theories are in turn corroborated by the successful prediction of future events. The argument is circular because observed regularity is claimed to generate the hypothesis of the uniformity of laws which is used in turn to predict future regularity. In fact, there is no necessity that a contingent event, once observed, will repeat itself unchanged! What is lacking is the logical justification for the inferred uniformity of laws. Riedl himself is well aware of the problem of induction, but he insists that it would be against common sense not to believe that the sun will rise again tomorrow. He is not willing to succumb to the impossibility of logical justification of induction, and predicts that the solar system will continue to exist for another 5 billion years (Riedl, 1985: 135). Whereas common sense must indeed rely on that hope (Rieppel, 1986d), Riedl cannot claim knowledge of this fact without running the risk of the "inductive turkey" characterized by Bertrand Russell (quoted in Chalmers, 1986: 16-17). After a certain period of time the turkey had concluded from everyday experience that there prevailed a uniformity of law that he would be fed each day at nine a.m. This law held until Christmas at which day he found himself dead instead of fed.

While Riedl is correct that we must hope for the sun to rise again tomorrow, he cannot, by his epistemology, justify any certainty of knowledge of that fact. It would be unfair to ask him for such assurance, however, since he only defends a hypothetical realism. All he claims is that our hope for the next morning's sunrise is safe enough to expect survival, because it has been tested an infinite number of times during the phylogenetic past of the human lineage. This brings us to that aspect of his theory which is the most exciting one in the present context. The hypothesis of the isomorphy of the structures of human cognition with some "objective" reality is claimed to be the "basis" (Riedl, 1986: 116) of evolutionary theory, a basis which furthermore cannot be doubted (Riedl, 1986: 31).

It is surprising to hit on yet another untestable assertion or first fact in a theory which claims to be based on "critical empiricism" (Riedl, 1986: 114). As Löw (1983) has pointed out, the test of the isomorphy of subjective structures of human cognition with an "extra-subjective" reality would require a position outside human thinking - obviously a logical as well as physical impossibility. Treating the "reality" of the world as a category of thought (Riedl, 1986: 114) will not help in the test of the hypothesis of isomorphy either. Riedl (1975, 1985, 1986; Riedl and Kaspar, 1980) proposed another test, however, and this is comparative biology. Riedl considers comparative biology to have demonstrated the existence of an uncontradicted hierarchical order in nature: "millions of species" are claimed to fall into an "uncontradicted" pattern of order in nature (Riedl, 1976: 42), which is exactly what has to be expected in view of the "fact" of evolution (Riedl and Kaspar, 1980: 41). Evolution, Riedl (1975: 5) concludes, entails mechanisms creating order in nature which can be discovered and described by scientists: there is a "striking" coincidence between the pattern of order in nature and the structure of human cognition (Riedl, 1975: 22). Comparative biology has established successful hypotheses of grouping which are corroborated each time new taxa are discovered. In view of the number of species known to modern biology, which (according to Riedl, 1976) all fall into a congruent pattern of order in nature, it must be concluded that the hypotheses of homologies formulated by comparative biology have successfully stood the test a great number of times and thus may be considered as highly corroborated - as isomorphic with the "real" world! Human thinking is apparently able to correctly grasp this pattern of order in nature, at least to a highly corroborated degree, and this ability is explained by the action of natural selection. As order in nature is postulated to have been created by evolution through natural selection, Riedl (1975, 1985, 1986; Riedl and Kaspar, 1980) concludes that the structures of human cognition, able to correctly perceive this order, must have been adapted to the structures of the "real" world by the same efficient cause. This argument entails a number of problems, amongst which the claim for the free interchangeability of *explanandum* and *explanans* (Riedl, 1985: 127), and the lawful relation of homology, figure prominently.

It was detailed above that the theory of evolution figures as first cause and absolute fact in Riedl's epistemology. From this fact the existence of order in nature is deduced and found to be corroborated by the results of comparative biology which disclose lawful regularity in the distribution of characters (homologies). The preceding chapters should have made amply clear that the flow of information must be, and indeed is, the other way around. Evolution has never been observed to create order in nature; what is observed is order in nature, and it is this observation which is explained on the theory of evolution (Brady, 1985). Evolution is not a first cause which must be "trusted" (Riedl, 1985: 287); instead, it is a theory explaining the observed pattern of order in nature. Riedl's (1975) concept of *burden* is not an *a priori* platform permitting the successful prediction of homologies; instead, it is an explanation for observed character congruence!

Riedl, finds the pattern of order in nature "uncontradicted", i.e. fully congruent and hence disclosing a law-like regularity in the relations of homology. This aspect has been discussed in some detail above. It was stated that the hierarchy of homology is not revealed by unbiased observation. Instead, the relation of homology must be intended, it is based on the hypothesis of hierarchical order in nature, and it requires methodological rules, i.e. operational criteria. It was shown above that a change of intention, of the working hypotheses or of methodological procedures may alter observed appearances. Quinarism was mentioned as a case in point. Riedl's theory of evolutionary epistemology would be hard pressed if asked to explain the popularity of quinarism early in the 19th century. If the isomorphy between order in nature and human thinking is due to natural selection, the structures of human cognition must be innate, as is indeed postulated by Riedl. But this cannot explain why it is that earlier naturalists could get excited about a pattern of order in nature which seems absurd if judged by modern standards. Similarly, it remains unclear how Riedl's (1976) theory, which postulates an uncontradicted pattern of order for millions of species as revealed by innate structures of cognition, can accommodate the different conceptions and delineations of species in folk taxonomies.

In fact, the isomorphy of human thinking with the hierarchy of homologies exists only within the limits of conceptual constraints which guide the observation of similarity. This comes as little surprise, since homology is a category of thought (Nelson, 1970: 378), denoting logical relations of form. If conceptualized within the hypothesis of a dichotomously subordinated hierarchy, the logical structure of the cladogram obtains. Dullemeijer (1980: 177) found "human thinking" to operate best in binary systems, and it was emphasized (chapter 6), that the hierarchically subordinated hierarchy is indeed the most economical way to organize observed similarity. The cladogram as such is atemporal and acausal, a conceptual construct which becomes a guide to ancestry only if the hypothesis of descent with modification is added to it (Dullemeijer, 1980), providing a causal explanation for observed regularity in the distribution of homologies (Brady, 1985; see also chapter 6).

The distribution of characters is not as regular as Riedl would have it, however. What we find instead is character incongruence, increasing with decreasing level of generality. The cladogram, or hierarchy of homologies, therefore cannot be claimed to match some supposedly "extra-subjective" reality of nature. Instead, we choose among several possible hierarchies the most parsimonious one, and use this as a guide to common ancestry in an evolutionary explanation of shared similarity. The groups which are thus interpreted as being descended from a common ancestor exist only within the limits of the application of the concept of homology, the inherent intentions, the inherent theory and the operational criteria, which guide the observation of shared similarity in the expectation of finding a hierarchical pattern of order in nature. True enough, it would be impossible to transcend mere multiplicity without guidance by the principle of homology (Brady, 1985: 124), but on the other hand, the pattern or order in nature may differ following different intentions, theories, and methodological rules.

Different 'Ways of Seeing'

The preceding chapters have treated pattern and process analysis as different 'ways of seeing', incompatible with each other but at the same time complementary to each other.

Pattern analysis focuses on the reconstruction of the hierarchy of "groups within groups". It seeks to demarcate groups from each other and their hierarchical level on the basis of shared similarity as conceptualized by the relation of homology. The relation of homology is a category of thought and as such ahistorical and acausal. The pattern specified by it emphasizes discontinuity between groups and the levels of the hierarchy they represent. Process analysis focuses on causal mechanisms of transformation. Transformation requires a minimal continuity of form, or, to put it the other way around: structures must be sufficiently close to give rise to a hypothesis of transformation linking one with the other. Process analysis is based on the principle of continuity, either of phenotypic effects (functionalism) or of genetic causes (structuralism). Pattern and process analysis are incompatible with each other as the first emphasizes discontinuity, while the other is based on the principle of continuity. The two 'ways of seeing' are complementary to each other, because pattern analysis is the only possible guide to common ancestry, providing process analysis with a direction, while process analysis is the only possibility to provide a causal explanation for pattern. This complementarity is also exemplified by the fact that the relation of homology can be represented either in a serial hierarchy, or in a subordinated hierarchy, whereby the two hierarchies can be translated into each other. The crucial point is that all continuity reveals itself to human perception (is conceptualized) as a series of discrete steps, however finely that series of steps may be graded; the complementary subordinated hierarchy will contain as many dichotomies as there are steps in the serial arrangement of the appearances.

It is interesting to find that the same two 'ways of seeing' obtain in Riedl's (1986) perspective, although they take on the nature of an inconsistency in the context of his evolutionary epistemology. On the one hand he postulates that all transitions from one type to another are gradual: the first mammals were hardly different from their reptilian ancestors, the first steamer was almost identical to a sailboat in appearance (Riedl, 1985: 118). The real world is one of continuous change and poorly defined boundaries, whereas the compartmentalization or subdivision of this world into concrete and discontinuous appearances is a product of conceptual thinking (Riedl, 1986: 160). On the other hand he postulates that discontinuity of things is a basic principle of nature (Riedl, 1986: 188), whereas the description or conceptualization of similarities imposes continuity of change as a result of linguistic constraints, constraints which are believed to be innate (Riedl, 1985: 192). One point borne out by this inconsistency of argumentation is that it throws into doubt the isomorphy of the structure of human cognition with some "extra-subjective" reality as discussed above. The other point is that Riedl obviously does not know what this world "really" is like: it is "complex, polymorphous, reticulate, typological" he asserts (Riedl, 1986: 140). This assertion is not convincing in face of the claim to have solved problems which haunted philosophy for over 25 centuries (Riedl, 1986: 15). More convincing is Riedl's (1986: 141) suggestion that we should start our approach to an understanding of the world around us with the admission that we cannot know anything. Following this suggestion we may find that we can look at this world in different ways, that we may see it differently, and that it will accordingly appear differently: static or dynamic, discontinuous versus continuous, as revealed by the taxic versus the transformational approach, specifying a subordinated versus a sequential hierarchy. There is no unifying access to the reality of the world around us, which is the intention of evolutionism. Instead, there are different 'ways of seeing' this world around us, each one with its merits and faults, each incomplete in itself, but complementary to one another.

EPILOGUE

At the basis of this book lies the conviction that there cannot be a unifying view of the world around us. All knowledge is conditioned by the structures of cognition, and all inter-subjective knowledge must be communicated by some language. Every experience presupposes the logical structure of notions, every notion is the expression of some intended relation, and every relation presupposes its opposite. Observation or perception entail a process of conceptualization if it is to transcend mere multiplicity in any meaningful way. Pattern and process are opposite 'ways of seeing', which provide in-*sight* into the structure and causality of the world around us. They are incompatible with each other, i.e. incomprehensible in simultaneous conjunction, as are the two sides of a coin that cannot be inspected simultaneously. They prove complementary to each other, however, as are *Anschauung* and *Kausalität*.

My argument started out from the dissatisfaction with a common intellectual attitude, namely to present evolution not as a theory (which it is), but as a fact (which supposedly cannot be doubted). As for anything else there are also two sides to the coin of evolution. One side is that of causality, i.e. the investigation of mechanisms of evolution in genetics, population biology, and ontogeny. The other side relates to the extrapolation of the operation of these mechanisms into the past, i.e. to phylogeny, explaining the order of nature we perceive. A very simple point to make is that it is impossible to observe the operation of evolutionary mechanisms in the distant past. We have not progressed very much beyond the basic point raised by F.-J. Pictet (1860: 243-244; translated by Hull, 1973: 144) in his review of Darwin's *On the Origin of species*: "But as I said in the beginning of this article, in Mr. Darwin's book there is a sudden leap by which the reader is asked to pass from the present study of facts to the most extreme theoretical consequences...". No one doubts the mechanisms of evolution as described by modern biology, acting at the populational level. The question is, however, whether, and if so how, these machanisms of evolution operated in the past to bring about the observed diversity of life. Only to ask the question whether causes in operation today account for processes in the past presupposes the principle of uniformity which cannot, in itself, be justified by experience but must constitute a logical and methodological *a priori* (Cassirer, 1973: 75-77; Rieppel, 1987a). Further, before we can answer the question whether the Darwinian mechanisms of evolution are necessary and sufficient to explain the historical diversity of organisms, or whether the theory of evolution must be amended in some or other way (Gould, 1980, 1982b; Eldredge, 1985), we must first have something to explain by these mechanisms, and this "something to be explained" (Brady, 1985) is the observed pattern of diversity!

All that can be observed is the pattern of character distribution in those organisms which are presumed to be the outcome of the hypothesized phylogenetic process. The fact that the analysis of character distribution began in pre-evolutionary times, and that the advent of evolutionary theory added next to nothing to the methods of pattern reconstruction (Russell, 1916,) documents that systematics is independent from evolutionary theory (Brady, 1985).

Pattern analysis must be free from assumptions about process (Rosen, Forey, Gardiner and Patterson, 1981), if the explanation of order in nature on the hypothesis of evolution is not to become circular (Brady, 1985). But pattern analysis cannot be free of all theory and any assumptions: it cannot be empirical in any sensualist sense. The reconstruction of pattern must start out from the expectation (hypothesis or conjecture) that a hierarchical order exists in nature, and that this can be recovered by the observation or conjecture of similarity or commonness of plan of construction. The perception or test of similarity must be guided by theoretical and methodological tools. Constitutional elements must be abstracted from a developmentally and functionally integrated whole, and these must be compared in abstract topological terms and in relation to some given frame of reference: this at least must hold as long as generative mechanisms of embryogenesis remain elusive for comparative purposes. It is important to note that any pattern recovered is not a matter of fact, but a matter of perception, that is of conceptualization: the hierarchy of homologies is an appearance of logical relations, an appearance which furthermore must be intended, which in other words obtains only within the accepted theoretical and methodological premises. If these are changed, the appearances change with them. When the intention or purpose of pattern reconstruction or classification changes, for instance from the storage and prediction of information concerning the distribution of topological relations of similarity to the storage and prediction of information concerning the distribution of characters relevant in the cultural context, the appearance of order in nature changes accordingly: tribal people conceptualize and delineate species different from trained biologists (Berlin, Breedlove and Raven, 1966; Brown, 1985).

Before anything else, classification has a purpose in itself: the ordering of the diversity of life into a system of maximally informative notions of groups. If some of these groups prove to be stable, we have reason to ask why that should be so. And alternatively: why do so many characters show incongruent distribution? A research program emerges, namely the quest for morphogenetic mechanisms imposing developmental and functional integration on the organization of matter. Similarity obtains from these generative mechanisms, and predictions of homologies based on the knowledge of the appropriate generative mechanisms will be successful as long as no change in the developmental program occurs.

Evolutionary theory on the other hand will always have to incorporate whatever hierarchical pattern is being recovered, and it will always prove complementary to it. That the theory of phylogeny is untestable is a consequence of its historical structure. More interesting is the claim that the mechanisms of evolution and the consequence of phylogeny cannot be doubted: this is an imperative which does not result from empirical observation, nor from logical necessity, but from historical contingency, i.e. from the role which evolutionary theory has come to play in the scientific community and in society at large. It is the claim of its indubitable status which renders evolutionary theory most interesting, because it raises the question of the motivation for that imperative!

The unity of type becomes explainable through the unity of descent on the hyphtesis of evolution. But as the unity of type is a concept embedded in and related to the methodology of pre-evolutionary biology, and thus is independent of evolutionary theory, the latter represents a superstructure, built over the observational basis of comparative biology. The question emerges addressing the motivation to add this superstructure, and its roots in social and political constellations in an ever-changing cultural context. Another question to answer is why that theory should have become so dominant in Western thought. Finally, recent developments in the field of epistemology and pattern reconstruction have led to the formulation of alternative models of evolutionay change, which have highlighted the implications of evolutionary theory for models of social change.

A process-oriented 'way of seeing' emphasizes serial arrangement and gradual change in small steps, each successive modification being continuously tested by natural selection and either preserved, and thus made the basis for further change, or else rejected: "Natural selection is daily and hourly scrutinizing throughout the world, every variation, even the slightest; rejecting what is bad, preserving and adding up all that is good; silently and insensibly working, whenever and wherever opportunity offers, at the improvement of each organic being in relation to its organic and inorganic conditions of life" as Darwin

(1859 [1959: 72]) put it. This functionalist approach seems to guarantee adaptation and progress: "And as natural selection works solely by and for the good of each being, all corporeal and mental endowments will tend to progress towards perfection". "There is grandeur in this view of life", is one of the closing phrases in Darwin's *Origin*, which permits a glimpse of the motivation underlying his defense of evolutionism. Karl R. Popper made the Darwinian 'view of life' the cornerstone of his epistemology: progress through trial and error! Science advances by the bold conjecture of new hypotheses, which have to be testable and potentially falsifiable. The conjecture of new hypotheses corresponds to the introduction of variation, the potential falsification is equivalent to a process of differential sorting. There results a concept of a steady progress of science which, in the aftermath of falsificationism, has proven a little too optimistic. Nevertheless, Popper has expanded his evolutionary perspective, both of nature and of *Objective Knowledge*, to a model for social change at large. His *Open Society* is one of continuous adaptation and progress, based on gradual change, social evolution taking place in small steps which have continuously to be tested. Should any social innovation prove advantageous, it is to be retained and made the basis for further change. If some change proves deleterious to society, it is to be reversed: the consequences of a reversal in cultural evolution are less damaging for society the smaller the initial change was. Nature, science and society: all progress by trial and error, i.e. by a continuous process of learning.

The taxic approach to the diversity of life emphasizes discontinuity in a subordinated hierarchical pattern instead. Gaps in the morphological continuum are explained by a structuralist understanding of organization. Form is created anew during each ontogeny, and submitted to natural selection as a *fait accompli*. It may either pass or fail. The concept of deviation permits radical change during short periods of time, but this range is random with respect to environmental (external) conditions. There is more to gain, but also more to lose in this type of game. Small wonder that one proponent of structuralism, S.J. Gould (in Gould and Eldredge, 1977: 146) commented: "It may also not be irrelevant to our personal preferences that one of us learnt his Marxism, literally at his daddy's knee".

This brief excursion into wider implications of the theory of evolution was only meant to show that the science of comparative biology, i.e. the reconstruction of pattern, still remains an important field of "observation and reflexion", as K.E. von Baer put it. Only knowledge of the conceptual and methodological background of pattern reconstruction permits the evaluation of different models of evolutionary change built over classifications. And this, in fact, is prerequisite for any understanding of motivations underlying the support of evolution in general, and various models of evolutionary change in particular, with all their cultural, social and political implications. Here begins another book, however.

REFERENCES

Adam, M., 1955. L'influence posthume. In: Pierre Gassendi 1592-1655. Sa Vie et son Oeuvre, pp. 157-182. Paris: A. Michel.

Adelman, H.B., 1966. Marcello Malpighi and the Evolution of Embryology. Ithaca, N.Y.: Cornell University Press.

Anderson, L., 1982. Charles Bonnet and the Order of the Known. Dordrecht and London: D. Reidl.

Agassiz, L., 1841. De la Succession et du Développement des Etres Organisés à la Surface du Globe Terrestre dans les Différens Ages de la Nature. Neuchâtel: Imprimerie Wolfrath.

Agassiz, L., 1844: Monographie des Poissons Fossiles du Vieux Grès Rouge ou Système Dévonien (Old Red Sandstone) des Iles Britanniques et de Russie. Soleure: Jent & Gassmann.

Agassiz, L., 1845: Iconographie des Coquilles Tertiaires réputées identiques avec les espèces vivantes ou dans différens terrains de l'époque tertiaire, accompagnée de la description des espèces nouvelles. Neuchâtel: H. Wolfrath.

Agassiz, L., 1859. An Essay on Classification. London: Longman, Brown, Green, Longmans and Roberts.

Agassiz, L., 1874. Evolution and the permanence of Type. In: D.L. Hull, Darwin and his Critics, pp. 430-445. Cambridge, Mass.: Harvard University Press.

Alberch, P., 1980. Ontogenesis and morphological diversification. Amer. Zool. **20**: 653-667.

Alberch, P., 1985. Problems with the interpreation of developmental sequences. Syst. Zool. **34**: 46-58.

Alberch, P. and E.A. Gale. 1985. A developmental analysis of an evolutionary trend: digital reduction in amphibians. Evolution **39**: 8-23.

Alberch, P., S.J. Gould, G.F. Oster and D.B. Wake, 1979. Size and shape in ontogeny and phylogeny. Paleobiology **5**: 296-317.

Andrews, P., 1987. Aspects of hominoid phylogeny. In: Patterson, C. (Ed.), Molecules and Morphology in Evolution: Conflict or Compromise?, pp. 23-53. Cambridge: Cambridge University Press.

Appel, T., 1987. The Cuvier-Geoffroy Debate. French Biology in the Decades before Darwin. Oxford: Oxford University Press.

Arnold, E.N., 1981. Estimating phylogenies at low taxonomic levels. Z. zool. Syst. Evolutionsforsch. **19**: 1-35.

Arthur, W., 1984. Mechanisms of Morphological Evolution. Chichester: John Wiley & Sons.

Augustinus, A., 1961. De Genesi ad Litteram Libri Duodecim, translated by C.J. Perl. Paderborn: Ferdinand Schöningh.

Ax, P., 1984. Das Phylogenetische System. Stuttgart: Gustav Fischer Verlag.

Ayala, F.J. and J.A. Kiger jr., 1980. Modern Genetics. Menlo Park Publ., CA: The Benjamin/Cummings Publ. Co.

Baer, K.E. von, 1828. Über Entwickelungsgeschichte der Thiere. Beobachtung und Reflexion, Theil I. Königsberg: Gebr. Bornträger.

Baer, K.E. von, 1873. Entwickelt sich die Larve der einfachen Ascidie in der ersten Zeit nach dem Typus der Wirbeltiere? Mém. Acad. Imp. Sci. St. Petersbourg **19**: 1-35.

Balss, H. (Ed.), 1943. Aristoteles biologische Schriften. München: Ernst Heimeran.

Baker, J.R., 1952. Abraham Trembley of Geneva. Scientist and Philosopher. 1710-1784. London: Edward Arnold & Co.

Beatty, J., 1982. What's in a word? Coming to terms in the Darwinian revolution. J. Hist. Biol. **15**: 215-239.

Beatty, J., 1985. Speaking of species: Darwin's strategy. In: Kohn, D. (Ed.), The Darwinian Heritage, pp. 265-281. Princeton, N.J.: Princeton University Press.

Bellairs, A.d'A., 1949. The anterior brain-case and interorbital septum of Sauropsida, with a consideration of the origin of snakes. J. Linn. Soc. (Zool.) **41**: 482-512.

Bellairs, A.d'A. and A.M. Kamal, 1981. The Chondrocranium and the development of the skull in Recent reptiles. In: Gans, C. and T.S. Parsons (Eds.), Biology of the Reptilia **11**: 1-163. London and New York: Academic Press.

Bendall, D.S. (Ed.), 1983. Evolution from Molecules to Man. Cambridge: Cambridge University Press.

Benton, M.J., 1985. Classification and phylogeny of the diapsid reptiles. Zool. J. Linn. Soc. **84**: 97-164.

Berlin, B., D.E. Breedlove and P.H. Raven, 1966. Folk taxonomies and biological classification. Science **154**: 273-274.

Bjerring, H.C., 1967. Does a homology exist between the basicranial muscle and the polar cartilage? Coll. Int. C.N.R.S. **163**: 223-267.

Bjerring, H.C., 1977. A contribution to structural analysis of the head of craniate animals. Zool. Scripta **6**: 127-183.

Blumenbach, J.F., 1781. Über den Bildungstrieb und das Zeugungsgeschäft. Göttingen: Johann Christian Dieterich.

Bock, W.J., 1959. Preadaptation and multiple evolutionay pathways. Evolution **13**: 194-211.

Bock, W.J., 1979. The synthetic explanation of macroevolutionary change - a reductionist approach. In: J.H. Schwartz and H.B. Rollins (Eds.), Models and Methodologies in Evolutionary Theory. Bull. Carnegie Mus. Nat. Hist. **13**: 20-69.

Bonde, N., 1977. Cladistic classification as applied to vertebrates. In: Hecht, M.K., P.C. Goody and B.M. Hecht (Eds.), Major Patterns in Vertebrate Evolution, pp. 741-804. New York: Plenum Press.

Bonnet, Ch., 1745. Traîté d'Insectologie; ou Observations sur les Pucerons, 2 vols. Paris: Durand Librairie.

Bonnet, Ch., 1760. Essai Analytique sur les Facultés de l'Ame. Copenhague: Frères Cl. and Ant. Philibert.

Bonnet, Ch., 1764. Contemplation de la Nature, 2 vols. Amsterdam: Marc-Michel Rey.

Bonnet, Ch., 1768. Considérations sur les Corps Organisés, 2nd ed., 2 vols. Amsterdam: Marc-Michel Rey.

Bonnet, Ch., 1769. La Palingénésie Philosophique, 2 vols. Geneva: C. Philibert and B. Chirol.

Bonnet, Ch., 1781. IID Mémoire sur la reproduction des membres de la salamandre aquatique, In: Oeuvres d'Histoire Naturelle et de Philosophie de Charles Bonnet, ed. in-8, vol. XI, pp. 114-115. Neuchâtel: Samuel Fauche.

Bordeu, T. de, 1751. Recherches Anatomiques sur la Position des Glandes et sur leur Action. Paris: G.-F. Quillau.

Bowler, P.J., 1971. Preformation and pre-existence in the seventeenth century. J. Hist. Biol. **4**: 221-244.

Bowler, P.J., 1974. Evolutionism in the Enlightenment. History of Science **12**: 159-183.

Brady, R.H., 1982: Dogma and doubt. Biol. J. Linn. Soc. **17**: 79-96.

Brady, R.H., 1983. Parsimony, hierarchy, and biological implications. In: Platnick, N.I. and A. Funk (Eds.), Advances in Cladistics **2**: 49-60. New York: Columbia University Press.

Brady, R.H., 1985. On the independence of systematics. Cladistics 1: 113-126.

Brock, G.T., 1932. Some developmental stages in the skulls of the geckos, *Lygodactylus capensis* and *Pachydactylus maculosa*, and their bearing on certain important problems in lacertilian craniology. S. Afr. J. Sci. **29**: 508-532.

Brongniart, A., 1800. Essai d'une classification naturelle des reptiles. Bull. Soc. Sci. Philomatique **35**: 81-82; **36**: 89-91.

Brown, C.H., 1985. Modes of subsistence and folk biological taxonomy. Current Anthropology **26**: 43-64.

Buffon, G., ca. 1855. Histoire des Animaux, first publ. in 1749. In: Flourens, P. (ed.), Oeuvres Complètes de Buffon, Vol. I. Paris: Garnier Frères.

Callot, E., 1965. La Philosophie de la Vie au XVIIIe siècle. Paris: Macel Rivière.

Carroll, R.L., 1985. A pleurosaur from the Lower Jurassic and the taxonomic position of the Sphenodontida. Palaeontographica A **189**: 1-28.

Carroll, R.L., 1987. Vertebrate Paleontology and Evolution. New York: W.H. Freeman & Co.

Carroll, R.L. and W.Lindsay, 1985. Cranial anatomy of the primitive reptile *Procolophon*. Can. J. Earth Sci. **22**: 1571-1587.

Cassirer, E., 1969. Philosophie und exakte Wissenschaft. Frankfurt a.M.: Vittorio Klostermann.

Cassirer, E., 1973. Die Philosophie der Aufklärung. Tübingen: J.C.B. Mohr (Paul Siebeck).

Chalmers. A.F., 1986. Wege der Wissenschaft. Heidelberg: Springer Verlag.

Chang, M.-M. and Y. Xiabo, 1983. Structure and phylogenetic significance of *Diabolichthys speratus* gen. et sp. nov., a new dipnoan-like form from the Lower Devonian of Eastern Yunnan, China. Proc. Linn. Soc. N.S.W. **107**: 171-184.

Charlesworth, B., R. Lande, and M. Slatkin, 1982. A neo-Darwinian commentary on macroevolution. Evolution **24**: 704-722.

Coleman, W., 1964. Georges Cuvier, Zoologist. Cambridge: Harvard University Press.

Copleston, F.C., 1976. Geschichte der Philosophie im Mittelalter. München: C.H. Beck.

Corsi, P. and P.J. Weindling., 1985. Darwinism in Germany, France, and Italy. In: Kohn, D. (Ed.), The Darwinian Heritage, pp. 683-729. Princeton, N.J.: Princeton University Press.

Cuvier, G., 1817. Le Règne Animal distribué d'après son Organisation. Paris: Deterville.

Cuvier, G., 1841. Histoire des Sciences Naturelles, complétée, rédigée, annotée et publiée par M. Magdeleine de Saint-Agy. Paris: Masson & Cie.

Darwin, C., 1859[1959]. On the Origin of Species. A Reprint of the First Edition, with a Foreword by Dr. C.D. Darlington, F.R.S. London: Watts & Co.

Darwin, Fr., 1887. The Life and Letters of Charles Darwin, 3 vols, 3rd ed. London: John Murray.

Daudin, F.M., 1803. Histoire Naturelle, Générale et Particulière, des Reptiles, vol. 7. Paris: F. Dufart.

Daudin, H., 1926a. Cuvier et Lamarck. Les Classes Zoologiques et l'Idée de Série Animale (1790-1830), 2 vols. Paris: Librairie Félix Alcan.

Daudin, H., 1926b. De Linné à Jussieu. Méthodes de la Classificationet Idée de Série en Botanique et en Zoologie (1740-1790). Paris: Librairie Félix Alcan.

deBeer, G.R., 1937. The Development of the Vertebrate Skull. Oxford: Oxford University Press.

deBeer, G.R., 1958. Embryos and Ancestors, 3rd ed. Oxford: Oxford University Press.

deBeer, G.R., 1954. *Archaeopteryx* and evolution. Advance. Sci. **42**: 1-11.

de Maillet, B., 1749. Telliamed, ou Entretiens d'un Philosophe Indien avec un Missionaire Francois. Basle: Libraires associés.

deQueiroz, K., 1985. The ontogenetic method for determining character polarity and its relevance to phylogenetic systematics. Syst. Zool. **34**: 280-299.

Dobzhansky, T., 1951. Genetics and the Origin of Species, 3rd ed. New York: Columbia University Press.

Dullemeijer, P., 1974. Concepts and Approaches in Animal Morphology. Assen: Van Gorcum & Co.

Dullemeijer, P., 1980. Functional morphology and evolutionary biology. Acta Biotheoretica **29**: 151-250.

Duméril, A., 1806. Analytische Zoologie. Aus dem Französischen, mit Zusätzen von L.F. Froriep. Weimar: Verlag des Landes-Industrie-Comptoirs.

Eldredge, N., 1971. The allopatric model and phylogeny in Paleozoic invertebrates. Evolution **25**: 156-167.

Eldredge. N., 1979. Alternative approaches to evolutionary theory. In: Schwartz, J.H. and H.B. Rollins (Eds.), Models and Methodologies in Evolutionary Theory. Bull. Carnegie Mus. Nat. Hist. **13**: 7-19.

Eldredge, N., 1985. The Unfinished Systhesis. Biological Hierarchies and Modern Evolutionary Thought. Oxford: Oxford University Press.

Eldredge, N. and J. Cracraft, 1980. Phylogenetic Patterns and the Evolutionary Process. New York: Columbia University Press.

Eldredge, N. and S.J Gould, 1972. Punctuated equilibria: an alternative to phyletic gradualism. In: Schopf, T.J.M. (Ed.), Models in Paleobiology, pp. 82-115. San Francisco: W.H. Freeman & Co.

Eldredge, N. and M.J. Novacek, 1985. Systematics and paleobiology. Paleobiology **11**: 65-74.

Eldredge, N., and S.N. Salthe, 1984. Hierarchy and evolution. In: Dawkins R. and M. Ridley (Eds.), Oxford Surveys in Evolution **1**: 182-206.

Farber, P.L., 1976. The type-concept in zoology during the first half of the nineteenth century. J. Hist. Biol. **9**: 93-119.

Feyerabend, P.K., 1981. Probleme des Empirismus. Braunschweig und Wiesbaden: Vieweg & Sohn.

Fink, W.L., 1982. The conceptual relationship between ontogeny and phylogeny. Paleobiology **8**: 254-264.

Fitzinger, L.I., 1826. Neue Classification der Reptilien nach Ihren natürlichen Verwandtschaften. Wien: Verlag J.G. Heubner.

Forey, P., 1982. Neontological analysis versus paleontological stories. In: Joysey, K.A. and A.E. Friday (Eds.), Problems of Phylogenetic Reconstruction, pp. 119-157. London and New York: Academic Press.

Forey, P., 1985. (Review of) Mechanisms of Morphological Evolution, by W. Arthur. Cladistics **1**: 396-399.

Forey, P., 1987. Relationships of lungfishes. J. Morph., Suppl. **1**: 75-91.

Frazzetta, T.H. 1962. A functional consideration of canial kinesis in lizards. J. Morph. **111**: 287-320.

Gans, C., 1960. Studies on amphisbaenids (Amphisbaenia, Reptilia). 1. A taxonomic revision of the Trogonophinae, and a functional interpretation of the amphisbaenid adaptive pattern. Bull. Amer. Mus. Nat. Hist. **119**: 129-204.

Gans, C., 1978. The characteristics and affinities of Amphisbaenia. Trans. zool. Soc., Lond. **34**: 347-416.

Gaffney, E.S., 1979. An introduction to the logic of phylogeny reconstruction. In: Cracraft, J. and N. Eldredge (Eds.), Phylogenetic Analysis and Paleontology, pp. 79-111. New York: Columbia University Press.

Gardiner, B.G., 1982. Tetrapod classification. Zool. J. Linn. Soc. **74**: 207-232.

Gaupp, E., 1900. Das Chondrocranium von *Lacerta agilis*. Ein Beitrag zum Verständnis des Amniotenschädels. Anat. Hefte **15**: 433-595.

Gaupp, E., 1902. Über die Ala temporalis des Säugerschädels und die Regio orbitalis einiger anderer Wirbeltierschädel. Anat. Hefte **19**: 155-230.

Gauthier, J. and K. Padian, 1985. Phylogenetic, functional, and aerodynamic analyses of the origin of birds and their flight. In: Hecht, M.K., J.H. Ostrom, G. Viohl and P. Wellnhofer (Eds.), The Beginnings of Birds, pp.185-197. Eichstätt: Jura-Museum.

Geoffroy Saint-Hilaire, E., 1824a. De l'aile operculaire ou auriculaire des Poissons, considérée comme un principal pivot, sur lequel doit rouler toute recherche de détermination des pièces composant le crâne des animaux. Mém. Mus. Hist. Nat., Paris, **11**: 420-444.

Geoffroy Saint-Hilaire, E., 1824b. Note complémentaire sur les prétendus osselets de l'ouie des poissons. Mém. Mus. Nat. Hist., Paris, **11**: 258-260.

Geoffroy Saint-Hilaire, E., 1824c: Sur une nouvelle détermination de quelques pièces mobiles chez la carpe. Mém. Mus. Nat. Hist.,**11**: 143-160.

Geoffroy Saint-Hilaire, E., 1825a. Sur quelques objections et remarques concernant l'aile operculaire ou auriculaire des Poissons. Mém. Mus. Nat. Hist., Paris, **12**: 13-17.

Geoffroy Saint-Hilaire, E., 1825b. Sur l'anatomie comparée des monstruosités animales par M. Serres. Rapport fait à l'Académie Royale des Sciences. Mém. Mus. Hist. Nat., Paris, **13**: 82-92.

Geoffroy Saint-Hilaire, E., 1825c. Anencéphales humains. Mém Mus. Nat. Hist., **12**: 257-292.

Geoffroy Saint-Hilaire, E., 1825d. Recherches sur l'organisation des gaviales. Mém. Mus. Nat. Hist., Paris, **12**: 97-155.

Geoffroy Saint-Hilaire, E., 1925e. Sur de nouveaux anencéphales humains, confirmant par l'autorité de leurs faits d'organisation la dernière théorie sur les monstres, et fournissant quelques élémens caractéristiques de plus et de nouvelles espèces au genre anencéphale. Mém. Mus. Hist. Nat., Paris, **12**: 233-256.

Geoffroy Saint-Hilaire, E., 1828a. Mémoire, où l'on se propose de rechercher dans quels rapports de structure organique et de parenté sont entre eux les animaux des âges historiques, et vivant actuellement, et les espèces antédiluviennes et perdues. Mém. Mus. Nat. Hist., Paris, **17**: 209-299.

Geoffroy Saint-Hilaire, E., 1828b. RAPPORT fait à l'Académie Royale des Sciences, sur un Mémoire de M. Roulin, ayant pour titre: Sur quelques changemens observés dans les animaux domestiques transportés de l'ancien monde dans le nouveau continent. Mém. Mus. Hist. Nat., Paris **17**: 201-208.

Geoffroy Saint-Hilaire, E., 1830. Principes de Philosophie Zoologique, discutés en Mars 1830, au Sein de l'Académie Royale des Sciences. Paris: Pichon et Dider.

Geoffroy Saint-Hilaire, E., 1833a. Le degré d'influence du monde ambiant pour modifier les formes animales; question intéressant l'origine des espèces téléosauriens et successivement celle des animaux de l'époque actuelle. Mém. Acad. R. Sci. Inst. France **12**: 63-92.

Geoffroy Saint-Hilaire, E., 1833b. Troisième mémoire des recherches faites dans les carières du calcaire oolithique de Caen, ayant donné lieu à la découverte de plusieurs beaux échantillons et de nouvelles espèces de téléosaures. Mém. Acad. R. Sci. Inst. France **12**: 44-61.

Ghiselin, M.T., 1966. An application of the theory of definitions to systematic principles. Syst. Zool. **15**: 127-130.

Ghiselin, M.T., 1969. The Triumph of the Darwinian Method. Berkeley: California University Press.

Ghiselin, M.T., 1974. A radical solution to the species problem. Syst. Zool. **23**: 536-554.

Ghiselin, M.T., 1981. Categories, life and thinking. Behav. Brain Sci. **4**: 269-283.

Ghiselin, M.T., 1987. Species concepts, individuality, and objectivity. Biology & Philosophy **2**: 127-143.

Gigon, O. (Ed.), 1982. Aristoteles. Einführungsschriften. München: DTV.

Gilson, E., 1955. History of Christian Philosophy in the Middle Ages. London: Shed and Ward.

Gingerich, P.D., 1979. The stratophenetic approach to phylogeny reconstruction in vertebrate paleontology. In: Cracraft, J. and N. Eldredge (Eds.), Phylogenetic Analysis and Paleontology, pp. 41-77. New York: Columbia University Press.

Gingerich, P.D., 1984. Punctuated equilibria - where is the evidence? Syst. Zool. **33**: 335-338.

Gingerich, P.D., 1985. Species in the fossil record: concepts, trends and transitions. Paleobiology **11**: 27-41.

Glass, B., 1959. Maupertuis, pioneer of genetics and evolution. In: Glass, B., O. Temkin and W.K. Straus jr., Forerunners of Darwin 1745-1859, pp. 84-112. Baltimore. Md.: The Johns Hopkins Press.

Goodman, M., M.R. Miyamoto and J. Czelusniak, 1987. Pattern and process in vertebrate phylogeny revealed by coevolution of molecules and morphologies. In: Patterson, C. (Ed.), Molecules and Morphology in Evolution: Conflict or Compromise?, pp. 141-176. Cambridge: Cambridge University Press.

Goodrich, E.S., 1916. On the classification of the Reptilia. Proc. Roy. Soc. Lond. **B 89**: 261-276.

Goodrich, E.S., 1930. Studies on the Structure and Development of Vertebrates. London: Macmillan & Co.

Goodwin, B.C., 1982. Development and evolution. J. theor. Biol. **97**: 43-55.

Goodwin, B.C., 1984a. Changing from an evolutionary to a generative paradigm in biology. In: Pollard, J.W. (Ed.), Evolutionary Theory; Paths into the Future, pp. 99-120. Chichester and New York: John Wiley & Sons.

Goodwin, B.C., 1984b. A relational field theory of reproduction and its evolutionary implications. In: Ho, M.-W. and P.T. Saunders (Eds.), Beyond Neo-Darwinism, pp. 219-21. London and New York: Academic Press.

Goodwin, B.C. and L.E.H. Trainor, 1983. The ontogeny and phylogeny of the pentadactyl limb. In: Goodwin, B.C., N. Holder and C.C. Wylie (Eds.), Development and Evolution, pp. 75-94. Cambridge: Cambridge University Press.

Gorniak, G.C., H.I.Rosenberg and C.Gans, 1982. Mastication in the tuatara, *Sphenodon punctatus* (Reptilia: Rhynchocephalia): structure and activity of the motor system. J. Morph. **171**: 321-353.

Gould, S.J., 1977. Ontogeny and Phylogeny. Cambridge, Mass.: Harvard University Press.

Gould, S.J., 1980. Is a new and general theory of evolution emerging? Paleobiology **6**: 119-130.

Gould, S.J., 1982a. The meaning of punctuated equilibria and its role in validating a hierarchical approach to macroevolution. In: Milkman, R. (Ed.), Perspectives on Evolution, pp. 83-104. Sunderland, Mass.: Sinauer Ass. Inc.

Gould, S.J., 1982b. Darwinism and the expansion of evolutionary theory. Science **216**: 380-387.

Gould, S.J., 1983. Irrelevance, submission, and partnership: the changing role of paleontology in Darwin's three centennials, and a modest proposal for macroevolution. In: Bendall, D.S. (Ed.), Evolution from molecules to men, pp. 347-366. Cambridge: Cambridge University Press.

Gould, S.J., 1984. Relationship of individual and group change. Hum. Dev. **27**: 233-239.

Gould, S.J., 1985. The paradox of the first tier: an agenda for paleobiology. Paleobiology **11**: 2-12.

Gould, S.J., 1988. The heart of terminology. Natural History (2) **1988**: 25-30.

Gould, S.J. and N. Eldredge, 1977. Punctuated equilibria: the tempo and mode of evolution reconsidered. Paleobiology **3**: 115-151.

Gould, S.J., D.M. Raup, J.J. Sepkoski jr. and D.S. Simberloff, 1977. The shape of evolution: a comparison of real and random clades. Paleobiology **3**: 23-40.

Gray, J.E., 1825. A synopsis of the genera of reptiles and amphibians, with a description of some new species. Ann. Philosoph., N.S. **10**: 193-217.

Greer, A.E., 1970. A subfamilial classification of scincid lizards. Bull. Mus. Comp. Zool. **139**: 151-184.

Gruber, H.E., 1985. Going to the limit: toward the construction of Darwin's theory (1832-1839). In: Kohn, D. (Ed.), The Darwinian Heritage, pp. 9-34. Princeton: Princeton University Press.

Haeckel, E., 1902. Natürliche Schöpfungsgeschichte, 2 vols., 10th edition. Berlin: Georg Reimer.

Hall, B.K., 1978. Developmental and Cellular Skeletal Biology. London and New York: Academic Press.

Haller, A.v., 1755. Dissertation sur les Parties Irritables et Sensitives des Animaux. Traduit du Latin par M. Tissot. Lausanne: Marc-Michel Bousquet.

Haller, A.v. 1758. Sur la Formation du Coeur dans le Poulet. Lausanne: Marc-Michel Bousquet.

Hanken, J., 1986. Morphological evolution. Evolution **40**: 443-444.

Harvey, W., 1628. An anatomical disquisition on the motion of the heart and blood in animals. In: Willis, R. (Ed.), 1965, The Works of William Harvey, pp. 1-86. New York: Johnson.

Harvey, W., 1651 (1981). Disputations Touching the Generation of Animals. Translated with introduction and notes by G. Whitteridge. London: Blackwell.

Heaton, M.J., 1979. Cranial anatomy of primitive captorhinid reptiles from the late Pennsylvanian and early Permian of Oklahoma and Texas. Bull. Oklahoma Geol. Survey 127: 1-84.

Hecht, M.K., 1985. The biological significance of *Archaeopteryx*. In: Hecht, M.K., J.H. Ostrom, G. Viohl and P. Wellnhofer (Eds.), The Beginnings of Birds, pp. 149-160. Eichstätt: Jura-Museum.

Hecht, M.K. and J.L. Edwards, 1977. The methodology of phylogenetic inference above the species level. In: Hecht, M.K., P.C. Goody and B.M. Hecht (Eds.), Major Patterns in Vertebrate Evolution, pp. 3-51. New York: Plenum Press.

Heise, H., 1981. Universals, particulars, and paradigms. The Behavioral and Brain Sciences 4: 289-290.

Hennig, W., 1967. Phylogenetic Systematics. Urbana: University of Illinois Press.

Hinchliffe, J.R., 1985. 'One, two, three' or 'two, three, four': an embryologist's view of the homologies of the digits and carpus of modern birds. In: Hecht, M.K., J.H. Ostrom, G. Viohl and P. Wellnhofer (Eds.) The Beginnings of Birds, pp. 141-174. Eichstätt: Jura-Museum.

Hinchliffe, R.J. and M.K. Hecht, 1984. Homology of the bird wing skeleton. In: Hecht, M.K., B. Wallace and G.T. Prance (Eds.), Evolutionary Biology 18: 21-39. New York: Plenum Press.

Hodge, M.J.S, 1985. Darwin as a lifelong generation theorist. In Kohn, D. (Ed.), The Darwinian Heritage, pp. 207-243. Princeton, N.J.: Princeton University Press.

Hoffheimer, M.H., 1982. Maupertuis and the eighteenth-century critique of preexistence. J. Hist. Biol. 15: 119-144.

Hoffstetter, R., 1955. Squamates de type moderne. In: Piveteau, J. (Ed.), Traîté de Paléontologie, vol. 5, pp. 606-662. Paris: Masson & Cie.

Hoffstetter, R., 1962. Revue de récentes acquisition concernant l'histoire et la systématique des squamates. Coll. Int. C.N.R.S. 104: 243-279.

Hoffstetter, R., 1968. (Review of) A contribution to the classification of snakes, by G. Underwood. Copeia 1968: 201-213.

Holder, N., 1983. Developmental constraints and the evolution of vertebrate digit patterns. J. theor. Biol. 104: 451-471.

Holmes, E.B., 1985. Are lungfishes the sistergroup of tetrapods? Biol. J. Linn. Soc. 25: 379-397.

Howes, G.B. and H.H. Swinnerton, 1901. On the development of the skeleton of the tuatara, *Sphenodon punctatus*, with remarks on the egg, on the hatching, and on the hatched young. Trans. zool. Soc., Lond. 16: 1-86.

Howgate, M.E., 1985. Problems of the osteology of *Archaeopteryx*. Is the Eichstätt specimen a distinct genus? In: Hecht, M.K., J.H. Ostrom, G. Viohl and P. Wellnhofer (Eds.) The Beginnings of Birds, pp. 105-112. Eichstätt: Jura-Museum.

Hughes, A.J. and D.M. Lambert, 1984. Functionalism, structuralism, and "ways of seeing". J. theor. Biol. 111: 787-800.

Hull, D.L., 1973. Darwin and His Critics. Cambridge, Mass.: Harvard University Press.

Hull, D.L., 1976. Are species really individuals? Syst. Zool. 25: 174-191.

Hull, D.L., 1978. A matter of individuality. Phil. Sci. 45: 335-360.

Hull, D.L., 1980. Individuality and selection. Ann. Rev. Ecol. Syst. 11: 311-332.

Hull, D.L., 1987. Genealogical actors in ecological roles. Biology and Philosophy 2: 168-184.

Imbach, R., 1981. Wilhelm Ockham. In: Höffe. O. (Ed.), Klassiker der Philosophie 1: 220-244. München: C.H. Beck.

Jacob, F., 1983. Molecular tinkering in evolution. In: Bendall, D.S. (Ed.), Evolution from Molecules to Men, pp. 131-144. Cambridge: Cambridge University Press.

Jarvik, E., 1981. Lungfishes, tetrapods, paleontology, and plesiomorphy. Syst. Zool. 30: 378-384.

Kawata, M., 1987. Units and passages: a view for evolutionary biology and ecology. Biology & Philosophy **2**: 415-434.

Kemp, T.S., 1982. Mammal-Like Reptiles and the Origin of Mammals. London and New York: Academic Press.

Kemp, T.S., 1988. Haemothermia or Archosauria? The interrelationships of mammals, birds and crocodiles. Zool. J. Linn. Soc. **92**: 67-104.

Kielmeyer, C.F., 1793. Über die Verhältnisse der organischen Kräfte untereinander in der Reihe der verschiedenen Organisationen, die Gesetze und Folgen dieser Verhältnisse. 2nd printing 1814. Tübingen: Christian Friedrich Osiander.

Kitcher, P., 1986. Bewitchment of the biologist. Nature **320**: 649-650.

Klotz, I.M., 1980. The N-ray affair. Scientific American **242**: 122-131.

Kluge, A.G., 1971. Concepts and principles of morphologic and functional studies. In: Waterman, A.J. *et al*. (Eds.), Chordate Structure and Function, p. 3-41. New York: Macmillan.

Kluge, A.G., 1985. Ontogeny and phylogenetic systematics. Cladistics **1**: 13-27.

Kluge, A.G. and R.E. Strauss, 1985. Ontogeny and Systematics. Ann. Rev. Ecol. Syst. **16**: 247-268.

Krohn, W., 1987. Francis Bacon. München: C.H. Beck.

Kuhn, T.S., 1973. Die Struktur wissenschaftlicher Revolutionen. Frankfurt a.M.: Suhrkamp stw.

Lambert, D. and T. Hughes, 1984. Misery of functionalism. Rivista di Biologia **77**: 477-492.

LaMettrie, J.O., 1865. L'Homme Machine, avec une introduction et des notes de J. Assézat. Paris: Frédéric Henry.

Lankaster, E.R., 1870. On the use of the term homology in modern zoology, and the distinction between homogenetic and homoplastic agreements. Ann. Mag. Nat. Hist. (4) **6**: 34-43.

Lauder, G.V., 1980. The role of the hyoid apparatus in the feeding mechanism of the coelacanth *Latimeria chalumnae*. Copeia **1980**: 1-9.

Lauder, G.V., 1982. Introduction. In: Russell, E.S., Form and Function, pp. xi-xlv. Chicago: The University of Chicago Press.

Lenoir, T., 1978. Generational factors in the origin of *Romantische Naturphilosophie*. J. Hist. Biol. **11**: 57-100.

Lepenies, W., 1978. Das Ende der Naturgeschichte. Frankfurt a.M.: Suhrkamp stw.

Levinton, J.S., 1983. Stasis in progress: the empirical basis of macroevolution. Ann. Rev. Ecol. Syst. **14**: 103-137.

Lewontin, R.C., 1972. Testing the theory of natural selection. Nature **236**: 181-182.

Lovejoy, A.O., 1936. The Great Chain of Being. Cambridge, Mass.: Harvard University Press.

Lovejoy, A.O., 1959a. Buffon and the problem of species. In: Glass, B., O. Temkin and W.K. Straus jr. (Eds.), Forerunners of Darwin 1745-1859, pp. 84-112. Baltimore. Md.: The Johns Hopkins University Press.

Lovejoy, A.O., 1959b. Recent criticism of the Darwinian theory of recapitulation: its grounds and its initiator. In: Glass, B., O. Temkin and W.K. Straus jr. (Eds.), Forerunners of Darwin 1745-1859, pp. 438-458. Baltimore. Md.: The Johns Hopkins University Press.

Lovtrup, S., 1977.The Phylogeny of Vertebrata. London and New York: John Wiley & Sons.

Lovtrup, S., 1978. On von Baerian and Haeckelian Recapitulation. Syst. Zool. **27**: 348-352.

Lovtrup, S., 1979. The evolutionary species: fact or fiction? Syst. Zool. **28**: 386-392.

Lovtrup, S., 1982. The four theories of evolution. Rivista Biol. **75**: 53-66.

Löw, R., 1983. Evolution und Erkenntnis - Tragweite und Grenzen der evolutionären Erkenntnistheorie in philosophischer Absicht. In: Lorenz, K. and F.W. Wuketits (Eds.), Die Evolution des Denkens, pp. 331-360. Zürich: R. Piper.

Lundberg, J.G., 1973. More on primitiveness, higher level phylogenies and ontogenetic transformations. Syst. Zool. **22**: 327-329.

Lurie, E., 1960. Louis Agassiz: A Life in Science. Chicago: The University of Chicago Press.

Marx, J., 1976. Charles Bonnet contre les Lumières. Studies on Voltaire and the Eighteenth Century **156** and **157**: 1-782.

Maslin, T.P., 1952. Morphological criteria of phylogenetic relationships. Syst. Zool. **1**: 49-70.

Maupertuis, P.-L. M. de, 1745. Vénus Physique. Paris: Aubier Montaigne, reprint 1980.

Maupertuis, P.-L. M. de, 1753. Sur la Génération des Animaux. In: Les Oeuvres de MR. de Maupertuis, vol. II. Hildesheim: Georg Olms Verlag, reprint 1974.

Maupertuis, P.-L. M. de, 1758. Système de la Nature. In: Les Oeuvres de MR. de Maupertuis, vol. II. Hildesheim: Georg Olms Verlag, reprint 1974.

Mayr, E., 1942. Systematics and the Origin of Species. Reprinted 1964, with a new preface by the author. New York: Dover Publ.

Mayr, E., 1963. Animal Species and Evolution. Cambridge, Mass.: The Belknap Press of Harvard University Press.

Mayr, E., 1969. Principles of Systematic Zoology. New York: McGraw-Hill.

Mayr, E., 1971. Populations, Species, and Evolution, 2nd printing. Cambridge, Mass.: The Belknap Press of Harvard University Press.

Mayr, E., 1974. Behavior programs and evolutionary strategies. Amer. Sci. **62**: 650-659.

Mayr, E., 1982. The Growth of Biological Thought. Cambridge, Mass.: The Belknap Press of Harvard University Press.

Mayr, E., 1987a. The ontological status of species: scientific progress and philosophical terminology. Biology & Philosophy **2**: 145-166.

Mayr, E., 1987b. Answers to these comments. Biology & Philosophy **2**: 212-220.

McKenna, M.C., 1987. Molecular and morphological analysis of high-level mammalian interrelationships. In: Patterson, C. (Ed.), Molecules and Morphology in Evolution: Conflict or Compromise?, pp. 55-93. Cambridge: Cambridge University Press.

Meckel, J.F., 1811. Beyträge zur vergleichenden Anatomie, vol. 2, part 1. Leipzig: Carl Heinrich Reclam.

Merriam, B., 1820. Versuch eines Systems der Amphibien. Marburg: Johann Christian Krieger.

Milner, A.R., 1988. The relationships and origin of living amphibians. In: Benton, M.J. (Ed.), The Phylogeny and Classification of the Tetrapods. The Systematics Association Special Volume **35A**: 59-102. Oxford: Clarendon Press.

Mishler, B.D. and R.N. Brandon, 1987. Individuality, pluralism, and the phylogeetic species concept. Biology & Philosophy **2**: 397-414.

Moore, J.R., 1985. Darwin of Down: the evolutionist as squarson-naturalist. In: Kohn, D. (Ed.), The Darwinian Heritage, pp. 435-481. Princeton, N.J.: Princeton University Press.

Moy-Thomas, J.A. and R.S. Miles, 1971. Palaeozoic Fishes. London: Chapman & Hall Ltd.

Nelson, G., 1970. Outline of a theory of comparative biology. Syst. Zool. **19**: 373-384.

Nelson, G., 1973. Negative gains and positive losses: a reply to J.G. Lundberg. Syst. Zool. **22**: 330.

Nelson, G., 1978. Ontogeny, phylogeny, paleontology, and the biogenetic law. Syst. Zool. **27**: 324-345.

Nelson, G., 1983. Reticulation in cladograms. In: Platnick, N.I. and V.A. Funk (Eds.), Advances in Cladistics **2**: 105-111. New York: Columbia University Press.

Nelson, G., 1985. Outgroups and ontogeny. Cladistics **1**: 29-45.

Nelson, G. and N.I. Platnick, 1981. Systematics and Biogeography, Cladistics and Vicariance. New York: Columbia University Press.

Nelson, G. and N.I. Platnick, 1984. Systematics and Evolution. In: Ho, M.-W. and P.T. Saunders (Eds.), Beyond Neo-Darwinism, pp. 143-158. London and New York: Academic Press.

Oppel, M., 1811. Die Ordnungen, Familien und Gattungen der Reptilien, als Prodrom einer Naturgeschichte derselben. München: Joseph Lindauer.

Ospovat, D., 1981. The Development of Darwin's Theory. Natural History, Natural Theology, and Natural Selection, 1838-1859. Cambridge: Cambridge University Press.

Oster, G. and P. Alberch, 1982. Evolution and bifurcation of developmental programs. Evolution **36**: 444-459.

Ostrom J.H., 1976. *Archaeopteryx* and the origin of birds. Biol. J. Linn. Soc. **8**: 91-182.

Ostrom J.H., 1985a. Introduction to *Archaeopteryx*. In: Hecht, M.K., J.H. Ostrom, G. Viohl and P. Wellnhofer (Eds.), The Beginnings of Birds, pp. 9-20. Eichstätt: Jura-Museum.

Ostrom J.H., 1985b. The meaning of *Archaeopteryx*. In: Hecht, M.K., J.H. Ostrom, G. Viohl and P. Wellnhofer (Eds.), The Beginnings of Birds, pp. 161-176. Eichstätt: Jura-Museum.

Outram, D., 1986. Uncertain legislator: Georges Cuvier's Laws of Nature in their intellectual context. J. Hist. Biol. **19**: 323-368.

Page, R.D.M., 1987. Graphs and generalized tracks: quantifying Croizat's panbiogeography. Syst. Zool. **36**: 1-17.

Panchen, A.L., 1982. The use of parsimony in testing phylogenetic hypotheses. Zool. J. Linn. Soc. **74**: 305-328.

Panchen, A.L. and T.R. Smithson, 1988. The relationships of the earliest tetrapods. In: Benton M.J. (Ed.), The Phylogeny and Classification of the Tetrapods. The Systematics Association Special Volume **35A**: 1-32. Oxford: Clarendon Press.

Patterson, C., 1977. The contribution of paleontology to teleostean phylogeny. In: Hecht, M.K., P.C. Goody and B.R. Hecht (Eds.), Major Patterns in Vertebrate Evolution, 579-643. New York: Plenum Press.

1978. Verifiability in systematics. Syst. Zool. **27**: 218-222.

1981a. Agassiz, Darwin, Huxley, and the fossil record of teleost fishes. t. Mus. (Nat. Hist.) Geol. **35**: 213-224.

1981b. Significance of fossils in determining evolutionary relationships. v. Ecol. Syst. **12**: 195-223.

1982. Morphological characters and homology. In Joysey, K.A. and A.E. Eds.), Problems of Phylogenetic Reconstruction, pp. 21-74. London and rk: Academic Press.

1983. How does phylogeny differ from ontogeny. In: Goodwin, B.C., N. and C.C. Wylie (Eds.), Development and Evolution, pp. 1-31. Cambridge: lge University Press.

1984. Preformation to punctuation. Review of: Evolution, the History of an Peter J. Bowler. Nature **311**: 587.

1987. Introduction. In: Patterson, C. (Ed.), Molecules and Morphology in Evolution: Conflict or Compromise?, pp. 1-22. Cambridge: Cambridge University Press.

Patterson, C. and D.E. Rosen, 1977. Review of ichthyodectiform and other Mesozoic fishes and the theory and practice of classifying fossils. Bull. Amer. Mus. Nat. Hist. **158**: 81-172.

Peyer, B., 1950. Goethe's Wirbeltheorie des Schädels. Vierteljahresschr. natf. Ges. Zürich, Suppl. 2/3: 1-129.

Piaget, J., 1967. Biologie und Erkenntnis, German translation 1974. Frankfurt a.M.: S. Fischer.

Pictet, F.-J., 1960. Sur l'origine de l'espèce. Archieves des Sciences Physiques et Naturelles, Bibliotheque Universielle, Revue Suisse et Etrangère, nouvelle Période **7**: 233-255.

Platnick, N.I., 1977. Cladograms, phylogenetic trees, and hypothesis testing. Syst. Zool. **26**: 438-442.

Popper, K.R., 1972. Objective Knowledge. An Evolutionary Approach. Oxford: The University Press.

Popper, K.R., 1976a. Logik der Forschung, 6th edition. Tübingen: J.C.B. Mohr (Paul Siebeck).

Popper, K.R., 1976b. Unended Quest. An Intellectual Autobiography. La Salle, Ill.: Open Court Publ. Co.

Popper, K.R. and J.C. Eccles, 1977. The Self and its Brain. An Argument for Interactionism. Heidelberg: Springer Verlag.

Poulsen, J.E. and E. Snorrason (Eds.), 1986. Nicolaus Steno 1638-1686. A reconsideration by Danish Scientists. Gentofte: Nordisk Insulinlaboratorium.

Presley, R. and F.L.D. Steel, 1976. On the homology of the alisphenoid. J. Anat. **121**: 441-459.

Raff, E.C., H.B. Diaz, H.D. Hoyle, J.A. Hutchens, M. Kimble, R.A. Raff, J.E. Rudolph and M.A. Subler, 1987. Origin of multiple gene families: are there both functional and regulatory constraints?. In: Raff, R.A. and E.C. Raff (Eds.), Development as an Evolutionary Process, pp. 203-238. New York: Alan R. Liss Publ.

Raff, R.A. and T.C. Kaufman, 1983. Embryos, Genes, and Evolution. New York: Macmillan Publ. Co.

Rachootin, S.P., 1987. Owen and Darwin reading a fossil: *Marauchenia* in a boney light. In: Kohn, D. (Ed.): The Darwinian Heritage, pp. 155-183. Princeton: Princeton University Press.

Rage, J.-C., 1982. L'histoire des serpents. Pour la Science, Avril **1982**: 16-27.

Réaumur, R.A.F. de, 1712. Sur les diverses réproductions qui se font dans les Ecrevisses. Mém. Acad. R. Sci., Paris **1712**: 226-245.

Réaumur, R.A.F. de, 1742. Mémoires pour servir à l'Histoire des Insectes, vol. VI. Paris: Imprimerie Royale.

Regnéll, H., 1967. Ancient Views on the Nature of Life. Lund: C.W.K. Gleerup.

Remane, A., 1952. Die Grundlagen des natürlichen Systems, der vergleichenden Anatomie und der Phylogenetik. Leipzig: Akademische Verlagsgesellschaft.

Riedl, R., 1975. Die Ordnung des Lebendigen. Systembedingungen der Evolution. Hamburg: Paul Parey.

Riedl, R., 1976. Die Strategie der Genesis. Zürich: R. Piper.

Riedl, R., 1982. The role of morphology in the theory of evolution. In: Grene, M. (Ed.), Dimensions of Darwinism, pp. 205-238. Cambridge: Cambridge University Press.

Riedl, R., 1985. Die Spaltung des Weltbildes: Biologische Grundlagen des Erklärens und Verstehens. Berlin and Hamburg: Paul Parey.

Riedl, R., 1986. Begriff und Welt: Biologische Grundlagen des Erkennens und Begreifens. Berlin and Hamburg: Paul Parey.

Riedl, R. and R. Kaspar, 1980. Biologie der Erkenntnis. Die stammesgeschichtlichen Grundlagen der Vernunft. Berlin and Hamburg: Paul Parey.

Rieppel, O., 1976. The homology of the laterosphenoid bone in snakes. Herpetologica **32**: 426-429.

Rieppel, O., 1979a. The classification of primitive snakes and the testability of phylogenetic theories. Biol. Zbl. **98**: 537-552.

Rieppel, O., 1979b. Ontogeny and the recognition of primitive character states. Z. zool. Syst. Evolutionsforsch. **17**: 57-61.

Rieppel, O., 1980. Homology, a deductive concept? Z. zool. Syst. Evolutionsforsch. **18**: 315-319.

Rieppel, O., 1982. The phylogenetic relationships of the genus *Acontophiops* Sternfeld (Sauria: Scincidae), with a note on mosaic evolution. Ann. Transvaal Mus. **33**: 241-257.

Rieppel, O., 1983a. Kladismus oder die Legende vom Stammbaum. Basle: Birkhäuser Verlag.

Rieppel, O., 1983b. Gradualismus und Punktualismus. Komplementäre Modelle evolutiven Wandels. Paläont. Z. **57**: 189-197.

Rieppel, O., 1984a. Atomism, transformism and the fossil record. Zool. J. Linn. Soc. **82**: 17-32.

Rieppel, O., 1984b. Miniaturization of the lizard skull: its functional and evolutionary implications. In: Ferguson, M.W.J. (Ed.), The Structure, Development and Evolution of Reptiles. Symp. zool. Soc. Lond. **52**: 503-520. London and New York: Academic Press.

Rieppel, O., 1984c. The upper temporal arcade of lizards: an ontogenetic problem. Rev. suisse Zool. **91**: 475-482.

Rieppel, O., 1984d. Können Fossilien die Evolution beweisen? Natur und Museum **114**: 69-74.

Rieppel, O., 1985a. The dream of Charles Bonnet (1720-1793). Gesnerus **42**: 359-367.

Rieppel, O., 1985b. Muster und Prozess: Komplementarität im biologischen Denken. Naturwissenschaften **72**: 337-342.

Rieppel, O., 1985c. Ontogeny and the hierarchy of types. Cladistics **1**: 234-246.

Rieppel, O., 1986a. Atomism, epigenesis, preformation and pre-existence: a clarification of terms and consequences. Biol. J. Linn. Soc. **28**: 331-341.

Rieppel, O., 1986b. Species are individuals: a review and critique of the argument. In: Hecht, M.K., B. Wallace and G.T. Prance (Eds.), Evolutionary Biology **20**: 283-317. New York: Plenum Press.

Rieppel, O., 1986c. Der Artbegriff im Werk des Genfer Naturphilosophen Charles Bonnet (1720-1793). Gesnerus **43**: 205-212.

Rieppel, O., 1986d. (Review of) Riedl, R., Die Spaltung des Weltbildes: Biologische Grundlagen des Erklärens und Verstehens. Cladistics **2**: 196-200.

Rieppel, O., 1987a. Punctuational thinking at odds with Leibniz - and Darwin. N. Jb. Geol. Paläont. Abh. **174**: 123-133.

Rieppel, O., 1987b. Pattern and process: the early classification of snakes. Biol. J. Linn. Soc. **31**: 405-420.

Rieppel, O., 1987c. The phylogenetic relationships within the Chamaeleonidae, with comments on some aspects of cladistic analysis. Zool. J. Linn. Soc. **89**: 41-62.

Rieppel, O., 1987d. (Review of) Riedl, R., Begriff und Welt: Biologische Grundlagen des Erkennens und Begreifens. Quart. Rev. Biol., **62**: 425.

Rieppel, O., 1988a. A review of the origin of snakes. In: Hecht, M.K., B. Wallace and G.T. Prance (Eds.), Evolutionary Biology **22**: 37-130. New York: Plenum Press.

Rieppel, O., 1988b. Louis Agassiz (1807-1973) and the reality of natural groups. Biology & Philosophy **3**: 29-47.

Riess, J., 1986. Locomotion, biophysics of swimming and phylogeny of the ichthyosaurs. Palaeontographica **A 192**: 93-155.

Robb, J., 1977. The Tuatara. Durham: Meadowfield Press Ltd.

Robinet, J.B., 1761-1766. De la Nature, 4 vols. Amsterdam: E. van Harrevelt.

Robinson, P.L., 1976. How *Sphenodon* and *Uromastyx* grow their teeth and use them. In: A.d'A. Bellairs and C.B. Cox (Eds.), Morphology and Biology of Reptiles. Linn. Soc. Symp. Ser. **3**: 43-64. London and New York: Academic Press.

Röd. W., 1978. Geschichte der Philosophie. Vol. 3. Philosophie der Neuzeit, 1: Von Francis Bacon bis Spinoza. München: C.H. Beck.

Röd, W., 1982. Descartes. Die Genese des Cartesischen Rationalismus. München: C.H. Beck.

Roe, S.A., 1981. Matter, Life and Generation. Eighteenth-century embryology and the Haller - Wolff debate. Cambridge: Cambridge University Press.

Roger, J., 1971. Les Sciences de la Vie dans la Pensée Française du XVIIIe Siècle, 2nd ed. Paris: Armand Collin.

Romero-Herrera, A.E., H. Lehmann, K.A. Joysey and A.E.Friday, 1978. On the evolution of myoglobin. Phil. Trans. R. Soc. Lond. **B 283**: 61-163.

Rosen, D.E., P.L. Forey, B.G. Gardiner and C. Patterson, 1981. Lungfishes. tetrapods, paleontology, and plesiomorphy. Bull. Amer. Mus. Nat. Hist. **167**:159-276.

Rosenberg, A., 1985. The Structure of Biological Science. Cambridge: Cambridge University Press.

Rosenberg, A., 1987. Why does the nature of species matter? Comments on Ghiselin and Mayr. Biology & Philosophy **2**: 192-197.

Roth, V.L., 1984. On homology. Biol. J. Linn Soc. **22**: 13-29.

Rudwick, M.J.S., 1972. The Meaning of Fossils. Episodes in the History of Paleontology. London: Macdonald.

Russell, E.S., 1916. Form and Function. Reprinted 1982, with a new introduction by George V. Lauder. Chicago: The University of Chicago Press.

Salthe, S.N., 1985. Evolving Hierarchical Systems. Their Structure and Representation. New York: Columbia University Press.

Savioz, R, 1948. Mémoires Autobiographiques de Charles Bonnet de Genève. Paris: Librairie Philosophique J. Vrin.

Schlegel, H., 1827. Erpetologische Nachrichten. Isis **20**: 281-294.

Schlegel, H., 1837. Essai sur la Physiognomie des Serpents, 2 vols. Amsterdam: Arnz & Co.

Schoch, R.M., 1986. Phylogeny Reconstruction in Paleontoloy. New York: Van Nostrand Reinhold Co.

Schultze, H.-P., 1969. Die Faltenzähne der rhipidistiiden Crossopterygier, der Tetrapoden und der Actinopterygier-Gattung *Lepisosteus,* nebst einer Beschreibung der Zahnstruktur von *Onychodus* (Struniiformer Crossopterygier). Palaeontograph. Ital. (N.S. 35) **65**: 63-137.

Schultze, H.-P., 1981. Hennig und der Ursprung der Tetrapoda. Paläont. Z. **55**: 71-86.

Schweber, S.S., 1985. The wider British context in Darwin's theorizing. In: Kohn, D. (Ed.), The Darwinian Heritage, pp. 35-69. Princeton, N.J.: Princeton University Press.

Serres, E., 1824. Explication du système nerveux des animaux invertébrés. Ann. Sci. Nat., Paris, **3**: 377-380.

Serres, E., 1827a. Recherches d'anatomie transcendante, sur les lois de l'organogénie appliquées à l'anatomie pathologique. Ann. Sci. Nat., Paris, **11**: 47-70.

Serres, E., 1827b. Théorie des formations organiques, ou recherches d'anatomie transcendante sur les lois d'organogénie, appliquées à l'anatomie pathologique. Ann. Sci. Nat., Paris, **12**: 82-143.

Serres, E., 1830. Anatomie transcendante. - Quatrième mémoire. Loi de symétrie et de conjugaison du système sanguin. Ann. Sci. Nat., Paris, **21**: 5-49.

Shubin, N.H. and P. Alberch, 1986. A morphogenetic approach to the origin and basic organization of the tetrapod limb. In: Hecht, M.K., B. Wallace and G.T. Prance (Eds.), Evolutionary Biology **20**: 319-387. New York: Plenum Press.

Shumaker, W., 1972. The Occult Sciences in the Renaissance: A Study in Intellectual Pattern. Berkeley: University of California Press.

Sibley, C.G. and J.E. Ahlquist, 1987. Avian phylogeny reconstructed from comparisons of genetic material, DNA. In: Patterson, C. (Ed.), Molecules and Morphology in Evolution: Conflict or Compromise?, pp. 95-121. Cambridge: Cambridge University Press.

Simpson, G.G., 1961. Principles of Animal Taxonomy. New York: Columbia University Press.

Sloan, P.R., 1985. Darwin's invertebrate progam, 1826-1836. In: Kohn, D. (Ed.), The Darwinian Heritage, pp. 71-120. Princeton, N.J.: Princeton University Press.

Sober, E., 1983. Parsimony in systematics: philosophical issues. Ann. Rev. Ecol. Syst. **14**: 335-357.

Sober, E., 1984. The Nature of Selection. Cambridge: MIT Press.

Sonntag, O., 1983. The Correspondence between Albrecht von Haller and Charles Bonnet. Bern: Hans Huber.

Sorabji, R.,1983. Time, Creation & the Continuum. London: Gerald Duckworth & Co.

Spaemann, R., P. Koslowski and R. Löw, 1984. Evolutionstheorie und menschliches Selbstverständnis. Weinheim: Acta Humaniora.

Stanley, S.M., 1975. A theory of evolution above the species level. Proc. Nat. Acad. Sci. USA **72**: 646-650.

Stanley, S.M., 1979. Macroevolution. Pattern and Process. San Francisco: W.H. Freeman & Co.

Starck, D., 1965. Embryologie, 2nd. ed. Stuttgart: G. Thieme.

Starck, D., 1979. Vergleichende Anatomie der Wirbeltiere, vol. 2. Berlin and Heildelberg: Springer-Verlag.

Stebbins, G.L., 1987. Species concepts: semantics and actual situations. Biology and Philosophy **2**: 198-203.

Stock, G.B. and S.V. Bryant, 1981. Studies of digit regeneration and their implications for theories of development and evolution of vertebrate limbs. J. exp. Zool. **216**: 423-433.

Swammerdam, J., 1752. Die Bibel der Natur. Leipzig: J.F. Gladitsch.

Tarsitano, S. and M.K. Hecht, 1982. A reconsideration of the reptilian relationship of *Archaeopteryx*. Zool. J. Linn. Soc. **69**: 149-182.

Taylor, E.H., 1972. Squamation in caecilians, with an atlas of scales. Univ. Kansas Sci. Bull. 49: 989-1164.

Thomson, K.S., 1966. Intracranial mobility in the coelacanth. Science **153**: 999-1000.

Thomson, K.S., 1967. Mechanisms of intracranial kinesis in fossil rhipidistian fishes (Crossopterygii) and their relatives. J. Linn. Soc. (Zool.) **46**: 223-253.

Thomson, K.S., 1970. Intracranial movement in the coelacanth fish *Latimeria chalumnae* Smith (Osteichthyes, Crossopterygi). Postilla **149**: 1-12.

Thomson, K.S., 1981. A radical look at fish-tetrapod relationships. Paleobiology **7**: 153-156.

Thomson, K.S., 1986. Natural Science in the 1830s: The link from Newton to Darwin. Amer. Sci. **74**: 397-399.

Trembley, A., 1943. Correspondence Inédite entre Réaumur et Abraham Trembley. Introduction par Emile Guyénot. Geneva: Georg & Cie

Vrba, E.S, 1983. Macroevolutionary trends: new perspectives on the roles of adaptation and incidental effect. Science **221**: 387-389.

Vrba, E.S. and N. Eldredge, 1984. Individuals, hierarchies and processes: towards a more complete evolutionary theory. Paleobiology **10**: 146-171.

Watrous, L.E. and Q.D. Wheeler, 1981. The out-group method of character analysis. Syst. Zool. **30**: 1-11.

Webster, G., 1984. The relations of natural forms. In: Ho, M.-W. and P.T. Saunders (Eds.), Beyond Neo-Darwinism, pp. 193-217. London and New York: Academic Press.

Webster, G. and B. Goodwin, 1982. The origin of species: a structuralist approach. J. Social Biol. Struct. **5**: 15-47.

Whetstone, K.N., 1983. Braincase of Mesozoic birds: I. New preparation of the "London" *Archaeopteryx*. J. Vert. Paleont. **2**: 439-452.

White, M.J.D., 1978. Modes of Speciation. San Francisco: W.H. Freeman & Co.

Whiteside, D.I., 1986. The head skeleton of the Rhaetian sphenodontid *Diphydontosaurus avonis* gen. et sp. nov. and the modernizing of a living fossil. Phil. Trans. R. Soc. Lond. **B 312**: 379-430.

Wiley, E.O., 1981. Phylogenetics. The Theory and Practice of Phylogenetic Systematics. New York and Chichester: John Wiley & Sons.

Williams, B., 1981. Descartes. Das Vorhaben der reinen philosophischen Untersuchung. Königstein: Athenäum Verlag.

Williamson, P.G., 1981. Paleontological documentation of speciation in Cenozoic molluscs from Turkana basin. Nature **293**: 437-43; 1982: Nature **296**: 608-612; 1983: Nature **304**: 659-663.

Willmann, R., 1979. Erhöhung der Evolutionsgeschwindigkeit bei Verringerung der Populationsgrösse. Umschau in Wissenschaft und Technik **79**: 451-453.

Willmann, R., 1985. Die Art in Raum und Zeit. Hamburg und Berlin. Paul Parey.

Wolff, K.F., 1764. Theorie von der Generation in zwo Abhandlungen. Berlin: Friedrich Wilhelm Birnstiel.

Wyss, A.R., M.J. Novacek and M.C. McKenna, 1987. Amino acid sequence versus morphological data and the interordinal relationships of mammals. Mol. Biol. Evol. **4**: 99-116.

Zangerl, R. and G.R. Case, 1976. *Cobelodus aculeatus* (Cope), an anacanthous shark from Pennsylvanian black shales of North America. Palaeontographica A **154**: 107-157.

INDEX